The Integrated Optimization of School Starting Times and Public Transport

Zur Erlangung des akademischen Grades eines

Doctor rerum naturalium (Dr.rer.nat.)

vom Fachbereich Mathematik

der Technischen Universität Darmstadt

genehmigte

Dissertation

von

Dipl.-Math. Armin Fügenschuh

aus Cuxhaven

Referent:	Prof. Dr. Alexander Martin
Korreferent:	Prof. Dr. Mirjam Dür

Tag der Einreichung:	20. April 2005
Tag der mündlichen Prüfung:	24. Juni 2005

Darmstadt, 2005

D 17

Bibliografische Information Der Deutschen Bibliothek

Die Deutsche Bibliothek verzeichnet diese Publikation in der Deutschen
Nationalbibliografie; detaillierte bibliografische Daten sind im Internet über
http://dnb.ddb.de abrufbar.

ISBN 3-8325-1037-0

Logos Verlag Berlin
Comeniushof, Gubener Str. 47,
10243 Berlin
Tel.: +49 030 42 85 10 90
Fax: +49 030 42 85 10 92
INTERNET: http://www.logos-verlag.de

Von drei Lehrern
habe ich diese Facetten der Mathematik gelernt:

Gerald Schmieder – die Schönheit,
Peter Pflug – die Schärfe,
Alexander Martin – die Nützlichkeit.

Ihnen sei diese Arbeit
in Dankbarkeit gewidmet.

Zusammenfassung

Die vorliegende Dissertation beschäftigt sich mit der Formulierung und Lösung eines mathematischen Modells zur integrierten Optimierung der Schulanfangszeiten und des Nahverkehrsangebots, kurz: IOSANA. Ziel ist es, die Schulanfangszeiten in die Optimierung der Abfahrtszeiten und der Umläufe von Bussen des öffentlichen Personennahverkehrs (ÖPNV) einzubeziehen, um auf diese Weise Einsparungen bei der Anzahl der eingesetzten Fahrzeuge gegenüber dem Status-Quo zu erzielen.

Das zu Grunde liegende mathematische Modell ist eine Erweiterung des bekannten Tourenplanungsproblems mit Zeitfenstern. Der neue Aspekt sind Kopplungsbedingungen innerhalb der Zeitfenster. Für diese neue Problemklasse führen wir die Bezeichnung Tourenplanungsproblem mit gekoppelten Zeitfenstern ein. IOSANA ist damit zugleich ein erster Anwendungsfall eines derartigen Modells mit gekoppelten Zeitfenstern. Instanzen dieses Problems sind theoretisch (im Sinne der Komplexitätstheorie) wie auch praktisch schwer zu lösen. Zur Lösung entwickeln und vergleichen wir verschiedene primale und duale Verfahren.

Auf der primalen Seite erweitern wir bekannte Heuristiken (sogenannte Greedy-Verfahren) zunächst um Aspekte der Kopplung von Zeitfenstern. Der nächste naheliegende Schritt ist die Parametrisierung der Auswahlfunktion, welche die Suchrichtung der Heuristik steuert. Zur Bestimmung "guter" Parameterwerte verwenden wir unter anderem Improving Hit-and-Run, ein zufallsbasiertes Verfahren der Globalen Optimierung.

Da es sich bei IOSANA um ein ganzzahliges lineares Programm handelt, lassen sich Branch-and-Cut Verfahren einsetzen, um untere Schranken für die Anzahl der eingesetzten Busse zu erhalten. Zur Verbesserung dieser Schranken untersuchen wir Möglichkeiten, die Modellformulierung zu verschärfen. Als besonders wirksam stellt sich die Reformulierung als Set-Partitioning-Problem heraus, wodurch jedoch die Zulässigkeit verloren geht.

Wir stellen die Resultate der in dieser Arbeit entwickelten Verfahren anhand verschiedener zufällig generierter und realer Instanzen vor. Reale Datensätze aus fünf deutschen Landkreisen wurden von unserem Projektpartner BPI-Consult zur Verfügung gestellt. Unsere Lösungen vergleichen wir mit dem Status-Quo in den jeweiligen Landkreisen.

Eine Einsparung von 10–30% der Fahrzeugflotte ist durch den Einsatz der integrierten Optimierung möglich. Jeder eingesparte Bus ist in etwa 30.000 Euro pro Jahr wert. Pro Landkreis errechnen sich so Einsparungen von bis zu 1,4 Mio. Euro jährlich. Rechnet man dieses auf alle 323 Landkreise Deutschlands hoch, so ergeben sich Einsparungen von bis zu 300 Mio. Euro an jährlichen Zuschüssen der Landkreise für die Schülerbeförderung.

Worauf warten wir noch?

Abstract

The presented doctoral thesis deals with the formulation and solution of a mathematical model for the integrated optimization of public transport and school starting times (IOSANA for short, from the German project acronym). It aims at an inclusion of the school starting times into the optimization of trips and schedules of buses, in order to achieve a reduction in the number of deployed vehicles compared to the status quo.

The underlying mathematical model turns out to be an extension of the well-known vehicle routing problem with time windows (VRPTW). The new aspect are coupling constraints among the time windows. For this new class of problems we propose the notion vehicle routing problem with coupled time windows (VRPCTW).

In this terminology IOSANA can be considered as the first application of a VRPCTW model. This problem is theoretically (in the sense of complexity theory) as well as practically (for large real-world instances) difficult to solve. For its solution we develop and compare different primal and dual algorithms.

On the primal side we first extend known heuristic methods (so-called greedy algorithms) to aspects of coupled time windows. The next related step is a parameterization of the scoring function which takes control of the search direction within the heuristic. For the computation of "good" parameters we use (among others) improving hit-and-run, a randomized method from the field of Global Optimization.

Since IOSANA is formulated as an integer linear program branch-and-cut algorithms can be used to derive lower bounds on the number of deployed vehicles. To improve the so derived lower bounds we present a bundle of different techniques (preprocessing, lifting, and cutting planes) to strengthen the LP relaxation. A reformulation as a set partitioning problem turns out to be very useful in this respect, but at the price of losing feasibility.

We present results of the different algorithms of this thesis using various randomly generated and real-world instances. Real-world data sets from five different German counties were provided by our project partner BPI-Consult. We compare our solutions to the respective status quo of the counties.

Reductions of 10–30% in the number of vehicles are possible by the integrated optimization. A single bus yields already savings of about 30,000 Euro per year. For the entire county this might add up to 1.4 million Euro per year. Extrapolated to all 323 of Germany's counties this would yield up to 300 million Euro of saved public money for the transport of pupils.

What are we waiting for?

Acknowledgements

First of all, I wish to thank Prof. Dr. Alexander Martin for giving me the graduate position at Darmstadt University of Technology. He supervised my first steps in the field of discrete and combinatorial optimization and gave me all the support as well as the freedom that was necessary to realize this work.

This work would not be possible without our collaborating industrial partner, the consultant company BPI-Consult (with dependences in Lörrach, Mainz, and Berlin). I am in particular grateful to Dr. Peter Stöveken (now at the Zentrum für integrierte Verkehrssysteme (ZIV), Darmstadt) whose inspiring talk at the Heureka 2002 conference in Karlsruhe laid the foundation for this work. Many thanks also to his former team at BPI-Consult for providing real-world data of the five counties presented in this work and many fruitful discussions: Dr. Christian Mehlert, Christiane Albrecht (now at Erfurter Verkehrsbetriebe (EVAG), Erfurt), Sonja Baldowski (now at Nordhessischer Verkehrs-Verbund (NVV), Kassel), Markus Hammrich (now at ZIV), Stefan Bohl, and Matthias Prick (now at ZIV).

I am grateful to ILOG CPLEX, Bad Homburg v.d.Höhe, for regularly extending our grant of the latest and greatest versions of CPLEX. It really is a great tool! Many thanks to Dr. Thorsten Koch (Konrad-Zuse-Zentrum für Informationstechnik (ZIB), Berlin) for developing his modeling language Zimpl, which made many things much easier than before. A special thank to Prof. Dr. Mirjam Dür (Darmstadt University of Technology) for pointing out the importance of improving hit-and-run, and to Benjamin Höfler for carefully proof-reading the entire manuscript.

Also many thanks to all my colleagues from the working groups AG7, AG8, AG10, and AG12 at our department in Darmstadt. It was always a pleasure to work with you. Also the travels to Hirschegg will be unforgettable for me.

Last but not least I want to thank my wife Marzena for her patience and my parents Ulrich and Wilma for their irresistible belief in me.

Darmstadt, April 2005   *Armin Fügenschuh*

Contents

Chapter 1

Introduction

Traffic peaks are peaks in cost. This is in particular true for rural counties, where public transport is focused on the demand of pupils. About half to two third of pupils in rural areas take a bus to get to school. Most of them are integrated in the public bus system, a minority is transferred by special purpose school buses. In any way the respective county in which the pupils live is responsible for their transfer. This in particular means that the county administration pays the fees. Since tax money is a scarce resource nowadays, the administration has great interest in reducing their expenses as much as possible. In Section 1.1 we take a closer look at the settings in a county. In Section 1.2 we present the main idea how to reduce the number of buses. This is the central part of the entire thesis. In Section 1.3 we discuss whether its consequences are in conformity with the German law and administrative regulations. In Section 1.4 we give an overview on the objectives and side constraints resulting from both, the main idea and the regulations. Finally in Section 1.5 we take a look at different approaches found in the literature about the efficient deployment of school buses in different parts of the world.

1.1 The Settings

The counties in Germany we mainly focus on are rural. By the term "rural" we refer to counties where the biggest city has no more than (say) 30,000 inhabitants, and around 150,000 people live in an area of about 1,000 square kilometers.

The optimization of public transport in rural areas is mostly an optimization of the traffic caused by pupils on their ways to school and back home, because they are a large, if not the largest group of customers. More than half of all pupils get to school by means of public transport, that is, about 5,000 – 15,000 pupils take the bus, the train, or the tram (in the sequel we simply speak of buses) to 30 – 120 different schools. The average way to school has a length of around 10 km, a few pupils travel even more than 30 km twice a day. Most pupils arrive at their school using a single bus, but some have to transfer one or even two times before they arrive at their school.

Beside the morning and afternoon peaks at the beginning and the end of the school, there is a much lower demand for public transport over the rest of the day (see Figure 1.1). The morning peak is caused by the fact that nearly all schools start at more or less the same time, in a narrow band around 8 o'clock. The afternoon peak is much lower since the schools release their pupils at different times.

Bus companies offer their service, the transport of customers from one bus stop to another, in so-called lines and passenger trips (or trips, for short). A line is a sequence of bus stops and a trip is a line together with arrival and departure times for each bus stop. Thus every line has at least one trip. The trips are then assigned to the buses, so that each bus has to serve some of the trips. This sequence of trips is called the schedule of a bus. Typical figures for an entire county are 50 to 220 buses (schedules) which are serving 200 to 1500 trips over the whole day, depending on the size of the county.

In general the bus companies are interested in deploying their buses in an efficient way. That means, it is desirable to reduce the size of the fleet and to reduce the deadhead trips between two passenger trips where no passengers are on board (and thus no income is generated). Until recently this in itself was a difficult optimization problem (attacked and solved in the seminal work of Löbel [52]). Today for a given set of trips (i.e., for fixed departure and arrival times) it can be solved routinely using the available standard software.

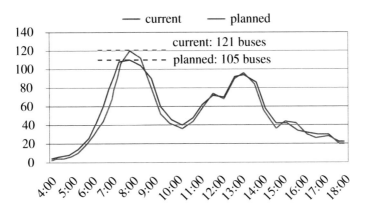

Figure 1.1: Deployed buses before and after the optimization.

1.2 The Central Idea

It is rather obvious that a significantly lower number of buses is needed, if the bus scheduling problem is solved together with the starting time problem, i.e., the simultaneous settlement of school and trip starting times, see Stöveken [69]. A small intuitive example is shown in Figure 1.2: If two schools start at the same time then two different buses are necessary to bring the pupils to their respective schools. If they start at different times then one and the same bus can first bring pupils to one school and then pupils to the other.

The consulting company *BPI-Consult*, a subsidiary of the Finnish *Jaakko Pöyry*, was one of the first to create a business model out of this observation. Between 1999 and 2004, BPI-Consult successfully consulted several counties, where they were able to find solutions (i.e., starting times and suitable bus schedules) which reduced the number of buses by 15 – 20%. Moreover as a part of their business concept, BPI-Consult does not only present a solution. Instead they accompany the whole embedding process, including negotiations with all participating groups (bus companies, pupils, parents, teachers, schools, and the county government). Within this process it is sometimes necessary to re-optimize the problem, when new, previously unknown constraints emerge. Interestingly, their income is proportional to the number of buses they saved in reality, not to their prediction at the beginning. However, their solutions are currently generated manually.

1.3 The Legal Framework

In this section we give an overview on the legal restriction for the optimization of school buses and school starting times.

Germany is divided into 16 states ("Länder") and sub-divided into 323 counties and 220 county-free cities ("Landkreise und kreisfreie Städte"). Within Germany's federal governmental system it is well-regulated which of their duties and tax-funded public services (p.e., waste collection, street preservation and rebuilding, or public transport) the states delegate to county administrations. For our special interest, we consider, as a representative example, the state's school law of Mecklenburg-Vorpommern (Mecklenburg-Western Pomerania) [67] and two administrative regulations [66], [39] from county Demmin and Mecklenburg-Vorpommern, respectively. The corresponding laws and regulations in other states and counties are similar.

Within the school law, the state declares its counties to be responsible for the transport of all their inhabitant pupils. The county has to carry out a public transport for all pupils attending public schools up to an age of 16, if they live too far away from school. The specification of what "too far away" means and how the public transport should be

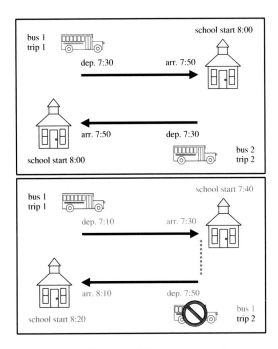

Figure 1.2: The central idea how to save buses.

organized, is at the discretion of the counties, assessing the endurance of pupils and the traffic safety of their ways to school.

Based on this document, county Demmin in Mecklenburg-Vorpommern specifies all missing details in an administrative regulation [66]. There, the "way to school" is defined as the shortest path between the pupils' home and the school which is responsible for the area they live in. (This might differ from the school the pupil actually is attending.) A school way is considered as unacceptable for pupils up to age 10, if it is longer than 2 km, and longer than 4 km for all others. Only those pupils are hence entitled to be transferred to school. Exceptions to this general rule are dangerous ways along highways without pedestrian walks, for instance, where even shorter ways are unacceptable. Also in these cases, the county pays for the transport cost of the pupils.

Moreover it is declared in [66] that the county is responsible for the organization of the transport by means of public transport. Usually, this is done by public buses and trains, in some rare cases by special school buses or rented cars. The deployment of all vehicles has to be done in such a way that pupils arrive not more than 60 minutes before school starts and do not have to wait more than 90 minutes after the end of school. If this is assured, "paying for transport" means that the county pays the fares for the pupils so that they can use one of the means of transport mentioned above. The starting and the ending time of schools should be settled in such a way that the vehicles (buses, trains, etc.) can be deployed in an economical way and traffic peaks are avoided. To achieve this, not all schools should start at the same time, rectified starting times are preferred. The transport companies and the school authorities are together responsible for a reconciliation of these times with public traffic.

By another administrative regulation of the state Mecklenburg-Vorpommern [39], schools are in general allowed to start between 7:30 and 8:30 a.m., exceptions regarding an earlier start are possible under certain circumstances. Here again, a rectification of school starting times is preferred to avoid traffic peaks. Thus, documents [66] and [39] constitute the legal foundation for the optimization problem we are aiming at.

1.4 The Problem

We are now heading towards a mathematical model for the problem outlined above. Turning the laws and administrative regulations into an optimization model, we identified the following variables, constraints and objectives. We focus on the following degrees of freedom (variables):

- The schedules of the buses,
- the starting times of the bus trips, and
- the starting times of the schools.

No other possible variables are issued, for example, planning the routes of the bus trips, or locating the bus stops. Moreover it is required that all pupils are using the same bus trips for their ways to school as at present. Within BPI-Consult's business model the decision which variables to focus on is mainly political. Changing the starting times of schools and bus trips causes already enough public opposition. For example, if some school starts at, say, 8:30 instead of 7:40, then the pupils leave home nearly one hour later, which might cause troubles for working parents. The same pupils then come home about one hour later, and therefore might no longer be able to attend a sports club. Reasons like this cause the changing of school starting times in a whole county at once to be a very delicate issue, and negotiating skills are needed to put through a new solution.

The decision variables are not independent from each other, they are coupled by the following constraints:

- The (legal) bounds on the school start (7:30 to 8:30 a.m.),

- lower and upper bounds on the waiting time for pupils at the school,

- bounds on the waiting time for pupils while transferring from one bus trip to another,

- bounds on the starting time of trips.

There are several conflicting goals that have to be modeled by the optimization. Bus companies, school authorities, pupils, parents, normal bus customers, and the county government are stake-holders in this "game", where each lobby group has its own interests, differing from the others. The county government wants to reduce their expenses for the pupils' fares. They emphasize a strong reduction of the number of deployed buses, and survey the compliance with legal restrictions. (BPI-Consult's goal is similar, because they want to maximize their profit.) Bus companies usually want to deploy their buses as efficiently as possible to be competitive with other companies sharing the same market. So they emphasize short deadhead trips and short idle times. Bus drivers are used to "their" schedules, which they in some cases consider as a "property" they do not want to give up. Pupils prefer buses which arrive at their school shortly before the start of the lessons. If they must change the bus, they don't want to wait too long for the connection. School authorities and parents would like to see the new school starting times close to the current ones so that they do not have to change their daily life. All other bus customers (who go by bus to work, for example) also don't like to change their routines, and hence prefer the new starting times of the buses close to the current ones. Summing up, we want to minimize the following:

- The total number of deployed buses,

- the time for driving deadhead-trips,

- the changes between current and planned bus schedules,

- the standing times of buses between two trips,

- the absolute change of the schools' starting times,

- the absolute change of the starting times of the bus trips,

- the waiting times for pupils at their schools, and

- the waiting times at a transfer bus stop.

Looking at these figures one can easily imagine that generating solutions manually is a difficult task, even if we only concentrate on the morning peak, which gives a small planning horizon from 5:00 – 9:00 a.m. Think of an average county consisting of about 2,000 bus stops, 100 schools, 250 bus trips, 10,000 pupils, and a bus company that deploys around 100 buses to serve all trips. Looking at this figures, it comes with no surprise that a human problem solver is easily mired by the huge amount of variables and constraints so that getting stuck in sub-optimal solutions is inevitable. Thus, the idea of an automatic planning tool was born. This planning tool was named IOSANA, which stands as an abbreviation for *Integrierte Optimierung der Schulanfangszeiten und des Nahverkehrs-Angebots* (integrated optimization of school starting times and public transport services).

1.5 Survey of the Literature on School Bus Optimization

The optimization of public transport is often based on a solution of a (multi-depot) vehicle scheduling problem, that is, the assignment of a set of vehicles to a set of bus trips with respect to several restrictions such that the number of deployed vehicles is minimized and the vehicles are used as efficiently as possible. Usually, the bus trips have fixed starting times, see Löbel [52] or Ginter, Kliewer, and Suhl [35], for instance.

A wide range of transport problems involving public bus transport and the organization of transporting pupils to their schools were already studied before. In this chapter we will take a closer look on some publications in this area. Afterwards we discuss how these results fit to the application we have at hand.

In [46] Keller and Müller describe a mixed integer programming approach for the determination of an optimal route of a (single) school bus in the street network around a school in Germany. In their approach, travel and waiting time of the pupils is taken into account. They formulate different objective functions considering the aspects of the pupils, the community, and the bus company. The model is solved using a standard mixed-integer programming solver to solve a problem instance with 97 pupils. It turns out that the resulting route is highly depending on the selected weights in the multicriteria objective function.

In [10] Bodin and Berman describe a procedure (heuristic algorithm) for a school bus routing and scheduling problem in Canada. The implementation of their algorithm is tested on data from two school districts and results in about a 20% savings in cost in one county and 600 additionally transported students with only one extra bus in the second. Their article is mainly focused on the preparation of the input data. They also point out that changing the school starting times in a district is crucial for an additonal saving of buses. Because of "political considerations" they determine the school starting times manually.

In [12] Braca et al. describe various issues related to the development of a computer software for the automatic solution of the school bus routing and scheduling problem in New York City (USA). The transport problem is divided into a morning and afternoon transport problem, where the morning problem is considered as more difficult, because the time windows are tighter. They formulate the problem in terms of two different integer programs: By the first one, a set of feasible routes is generated, and by the other a subset is selected which is sufficient to serve all pupils (set covering).

In [11] Bowerman et al. present a multi-objective model for urban school bus routing problem arising in Ontario (Canada). An additional degree of freedom is the integration of the bus stop planning problem, that is locating bus stops and assigning pupils to them is a part of the whole. The output performance of the model is measured by multiple objectives in terms of efficiency (cost, number of buses), effectiveness (how well is the demand satisfied), and equity (length and load balance among all trips, walking distances to bus stops). For the solution of the model with its related sub-problems, they suggest a sequence of several heuristics, where the results of each heuristic are merged into the final output.

In [26] Desaulniers et al. consider the urban bus scheduling problem (also in Canada), and formulate it as a multi-depot vehicle scheduling problem with time windows. This latter problem consists of scheduling a fleet of vehicles (buses) to cover a set of tasks (trips) at minimum cost. Each task is restricted to begin within a prescribed time interval and vehicles are supplied by different depots. The problem is formulated as an integer nonlinear multi-commodity network flow model with time variables and is solved using a column generation approach embedded in a branch-and-bound framework.

In [21] Corberan et al. address the problem of routing school buses in rural areas in Spain. Their approach is based on a node routing model with multiple objectives that arise from two conflicting goals, costs and quality of service. Their solution algorithm uses construction and improvement heuristics and combines different solutions within an evolutionary framework.

In [9] Bierlaire, Liebling, and Spada formulate a multi-objective model for school bus routing problems in Switzerland. They present a nonlinear mixed-integer programming formulation of the problem. For the generation of feasible solutions they implement and compare three different heuristic approaches (local search, tabu search, and simulated annealing), where the simulated annealing strategy performs better than the other two.

So far, none of the presented models completely fits our problem, mainly for some or all of the following reasons. The school starting times are fixed and cannot be changed to save buses. In all modeling approaches, pupils are always transported directly to school, and changing the bus is not allowed. Locating bus stops, designing routes (trips) and assigning pupils to routes is sometimes part of the optimization, but for us these are input figures. Finally, scheduling drivers is not an issue for us: Since our time horizon is small (from 5:00 to 9:00 a.m.), planning of breaks is not needed.

Chapter 2

Mathematical Preliminaries

To understand the models and algorithms presented in this thesis we provide in this chapter the necessary mathematical foundation. The fundamentals of graph theory are given in Section 2.1. The models presented later are integer linear programs, the definitions and results of this area are given in Section 2.2. For an introduction to complexity theory (i.e., the theory of P and NP) we refer the reader to the books of Garey and Johnson [34] or Nemhauser and Wolsey [59].

2.1 Graphs and Shortest Paths

In this section we summarize some basic notions from graph theory that are frequently used in the sequel. For a more detailed introduction see Clark and Holton [17], for instance.

A *graph* G is a tuple (V, E), consisting of a finite, nonempty set V and a set $E \subseteq \{\{v_1, v_2\} : v_1, v_2 \in V, v_1 \neq v_2\}$. The elements of V are called *vertices* or *nodes*, the elements of E are called *edges*. Node $v \in V$ is *adjacent* to node $w \in V$ if there is an edge $\{v, w\} \in E$. A *directed graph*, or *digraph*, is a tuple (V, A), where V is a finite, nonempty set of nodes, and a set $A \subseteq \{(v_1, v_2) : v_1, v_2 \in V, v_1 \neq v_2\}$ of so-called *arcs*. (In the sequel we will often call digraphs also graphs and omit the "di-" prefix. When talking about a "graph having arcs", a digraph is meant.) Remark that to graph (V, E) there is an associated digraph (V, A) with $A := \{(v, w), (w, v) : \{v, w\} \in E\}$. We say that $a \in A$ is *incident* to $v \in V$ if there exists $w \in V$ such that either $a = (v, w) \in A$ or $a = (w, v) \in A$. In the first case, $a = (v, w)$, we say v is a *start point* of a, in the latter case, $a = (w, v)$, v is called *end point* of a. A graph is called *complete* if it contains all possible edges, that is, $E := \{\{v, w\} \in V \times V : v \neq w\}$. Similar, a digraph is called *complete* if it contains all possible arcs, that is, $A := \{(v, w) \in V \times V : v \neq w\}$. Note that by our definition of graphs (digraphs) so-called parallel edges (arcs) are excluded, that is, there is at most one edge between nodes v and w in G (or at most one arc from v to w in the case G is a digraph).

A notion we will meet at several occasions in this thesis is that of a path, which is a special kind of walk. A v_0-v_N-walk in G of *length* N is a sequence of nodes $(v_0, v_1, \ldots, v_N)$, $N \geq 1$ with $(v_{i-1}, v_i) \in A$ for all $i = 1, \ldots, N$. Node v_0 is called *origin*, node v_N *destination*, and all other nodes $v_1, \ldots, v_{N-1}$ are *inner nodes*. Note that a walk can also be represented by its arc sequence $(a_1, a_2, \ldots, a_N)$, where $a_i = (v_{i-1}, v_i)$ for all $i = 1, \ldots, N$. If the nodes of a walk are pairwise distinct, it is called a *path*. A walk of length N with $v_0 = v_N$ is called a *closed walk*. A walk which is closed and a path is called a *cycle*.

A function $\omega : A \mapsto \mathbb{Q}_+$ is called *weight function*, and $\omega(a)$ is the *weight* of arc $a \in A$. The weight of a path $(a_1, a_2, \ldots, a_N)$ is defined as $\omega(a_1, a_2, \ldots, a_N) := \sum_{i=1}^{N} \omega(a_i)$. The *shortest path problem* can be stated as follows: Given a graph $G = (V, A)$, a weight function ω, an origin $v \in V$, and a destination $w \in V$, find a v-w-path of minimum weight. A path of minimum weight is also called *shortest path*. For the computation of a shortest path in a given graph, several algorithms are known from literature. We present here Dijkstra's algorithm, which in fact solves a more general problem: It finds shortest paths from node v to all other nodes.

dijkstra(G, ω, v)

Input:	graph $G = (V, A)$, weight function ω, origin node v
(1)	Let $g(v) := 0$, and $g(w) := \infty$ for all $w \in V, w \neq v$
(2)	**For All** $w \in V$ **Do**
(3)	**If** $(v, w) \in A$ **Then**
(4)	Let $p(w) := v$
(5)	**Else**
(6)	Let $p(w) := w$
(7)	**End If**
(8)	**End Do**
(9)	Let $T := V$
(10)	**Until** $T = \emptyset$ **Do**
(11)	Choose $w \in \operatorname{argmin}\{g(u) : u \in T\}$
(12)	**For All** arcs $a = (w, u)$ with $u \in T$ and $g(u) > g(w) + \omega(a)$ **Do**
(13)	Let $g(u) := g(w) + \omega(a)$
(14)	Let $p(u) := w$
(15)	**End Do**
(16)	Let $T := T \backslash \{w\}$
(17)	**End Do**
Output:	shortest path function $g : V \to \mathbb{Q}_+ \cup \{\infty\}$

Dijkstra's algorithm, named after the inventor and first presented in [28], constructs a function $g : V \mapsto \mathbb{Q}_+ \cup \{\infty\}$ such that for an arbitrary $w \in V$, $g(w)$ is the weight of a shortest path from v to w. If $g(w) = \infty$, then there's no path from v to w at all.

We set $\infty + x = \infty$ for an arbitrary real number $x \in \mathbb{Q}$, and $\infty + \infty = \infty$. It can be shown by induction over the set of nodes that Dijkstra's algorithm is correct and of order $O(n^2)$ (see Nemhauser and Wolsey [59] for instance).

A shortest path from v to w can be generated by the function $p : V \mapsto V$ that was constructed during the course of the algorithms. Here $p(w)$ is the node before w on a v-w-path. Thus, an entire shortest path from v to w is given by recursively applying function p to obtain an integer $N \leq |V|$ with $p^N(w) = v$ and $(v, p^{N-1}(w), \ldots, p^2(w), p(w), w)$ as a shortest v-w-path.

Dijkstra's algorithm is a famous example for an algorithm having polynomial-time complexity. Above we mentioned the Hamiltonian circuit problem, which in contrast is a typical NP-complete problem. An instance of the problem is specified by a graph $G = (V, E)$ or $G = (V, A)$ (for the undirected or the directed variant of the problem, respectively). The question is whether a Hamiltonian circuit in G exists, i.e., a cycle in G that visits every vertex exactly once. A proof for the NP-completeness of this problem can be found in Garey and Johnson [34].

2.2 Mixed-Integer Programming

The study and solution of linear integer programs lies at the heart of discrete optimization. Various problems in science, technology, business, and society can be modeled as linear integer programming problems and their number is tremendous and still increasing. Among the currently most successful methods are linear programming (LP, for short) based branch-and-bound algorithms where the underlying linear programs are possibly strengthened by cutting planes. For example, most commercial integer programming solvers, see Fourer [30], or special purpose codes for problems like the traveling salesman problem (see Section 3.1) are based on this method.

A *mixed integer program* (MIP) is a system of the following form:

$$\begin{aligned}
z_{\text{MIP}} = \quad \min \quad & c^T x \\
\text{subject to} \quad & Ax \leq b \\
& \underline{x} \leq x \leq \overline{x} \\
& x \in \mathbb{Z}^N \times \mathbb{Q}^C,
\end{aligned} \tag{2.1}$$

where $A \in \mathbb{Q}^{M \times (N \cup C)}$, $c \in \mathbb{Q}^{N \cup C}$, $b \in \mathbb{Q}^M$. Here, M, N and C are non-empty, finite sets with N and C disjoint. Without loss of generality, we may assume that the elements of N and C are represented by numbers, i.e., $N = \{1, \ldots, p\}$ and $C = \{p+1, \ldots, n\}$. The vectors $\underline{x} \in (\mathbb{Q} \cup \{-\infty\})^{N \cup C}$, $\overline{x} \in (\mathbb{Q} \cup \{\infty\})^{N \cup C}$ are called *lower* and *upper bounds* on x, respectively. A variable $x_j, j \in N \cup C$, is *unbounded from below (above)*, if $l_j = -\infty$ $(u_j = \infty)$. An integer variable $x_j \in \mathbb{Z}$ with $\underline{x}_j = 0$ and $\overline{x}_j = 1$ is called *binary*. If $N = \emptyset$ then (2.1) is called *linear program* or *LP*. If $C = \emptyset$ then (2.1) is called *integer program* or *IP*. We call $c^T x$ the *objective function* and $Ax \leq b, \underline{x} \leq x \leq \overline{x}, x \in \mathbb{Z}^N \times \mathbb{Q}^C$ the *side constraints* (or *constraints*, for short). The constraints $\underline{x} \leq x \leq \overline{x}$ are called *trivial constraints* or *bounds*. A solution vector x that satisfies all side constraints is called *feasible solution*.

From a complexity point of view mixed integer programming problems belong to the class of NP-hard problems (see Garey and Johnson [34], for example) which makes it unlikely that efficient, i.e., polynomial time, algorithms for their solution exist. The route one commonly follows to solve an NP-hard problem like (2.1) to optimality is to attack it from two sides. First, we consider the primal side and try to find some good feasible solution in order to determine an upper bound. Next, we consider the dual side and determine a lower bound on the objective function by relaxing the problem. The common basic idea of relaxation methods is to get rid of some part of the problem that causes difficulties. The methods differ in their choice of which part to delete and in the way to reintroduce the deleted part. The most commonly used approach is to relax the integrality constraints to obtain a linear program, because linear programs can efficiently be solved using Dantzig's simplex algorithm or interior point methods (for an introduction to linear programming see Nemhauser and Wolsey [59], or Bertsimas and Tsitsiklis [8], for instance). The integrality conditions are then reintroduced by iteratively adding cutting planes. If we are lucky the best lower and upper bounds coincide and the problem is solved. If not, we have to resort to some enumeration scheme, and the one that is most frequently used in this context is the branch-and-bound method.

2.2.1   Preprocessing

Preprocessing is the name for a whole bundle of simple inspection techniques for mixed-integer programs that are reported to decrease the solution time within a branch-and-bound or branch-and-cut framework (for a definition of these two concepts see Section 2.2.3 below). Here we concentrate on *bounds strengthening* where we exploit the bounds on the variables to detect so-called forcing and dominated rows. See for example Martin [54] for a survey of several other techniques.

Consider some row i and let

$$
\begin{aligned}
L_i &= \sum_{j \in P_i} a_{ij}\underline{x}_j + \sum_{j \in N_i} a_{ij}\overline{x}_j, \\
U_i &= \sum_{j \in P_i} a_{ij}\overline{x}_j + \sum_{j \in N_i} a_{ij}\underline{x}_j
\end{aligned}
\tag{2.2}
$$

where $P_i = \{j : a_{ij} > 0\}$ and $N_i = \{j : a_{ij} < 0\}$. Obviously, $L_i \leq \sum_{j=1}^n a_{ij}x_j \leq U_i$. The following cases might come up. An inequality i is an *infeasible row* if $L_i > b_i$. In this case the entire problem is infeasible. An inequality i is a *forcing row* if $L_i = b_i$. In this case all variables in P_i can be fixed to their lower bound and all variables in N_i to their upper bound. Row i can be deleted afterwards. An inequality i is a *redundant row* if $U_i < b_i$. In this case i can be removed.

This row bound analysis can also be used to strengthen the lower and upper bounds of the variables. Compute for each variable x_j and each inequality i

$$u_{ij} = \begin{cases} (b_i - L_i)/a_{ij} + \underline{x}_j, & \text{if } a_{ij} > 0 \\ (L_i - U_i)/a_{ij} + \underline{x}_j, & \text{if } a_{ij} < 0 \end{cases}$$

$$l_{ij} = \begin{cases} (L_i - U_i)/a_{ij} + \overline{x}_j, & \text{if } a_{ij} > 0 \\ (b_i - L_i)/a_{ij} + \overline{x}_j, & \text{if } a_{ij} < 0. \end{cases}$$

(2.3)

Let $u_j = \min_i u_{ij}$ and $l_j = \max_i l_{ij}$. If $u_j \leq \overline{x}_j$ and $l_j \geq \underline{x}_j$ we speak of an *implied free variable*. The simplex method might benefit from not updating the bounds but treating variable x_j as a free variable (note that setting the bounds of x_j to $-\infty$ and $+\infty$ will not change the feasible region). Free variables will commonly be in the basis and are thus useful in finding a starting basis. For mixed integer programs however, it is better in general to update the bounds by setting $\overline{x}_j = \min\{\overline{x}_j, u_j\}$ and $\underline{x}_j = \max\{\underline{x}_j, l_j\}$, because the search region of the variable within an enumeration scheme is reduced. In case x_j is an integer (or binary) variable we round u_j down to the next integer and l_j up to the next integer. As an example consider the following inequality (also taken from Martin [54]):

$$-45x_6 - 45x_{30} - 79x_{54} - 53x_{78} - 53x_{102} - 670x_{126} \leq -443$$

(2.4)

Since all variables are binary we get $L_i = -945$ and $U_i = 0$. For $j = 126$ we obtain $l_{ij} = (-443 + 945)/ -670 + 1 = 0.26$. After rounding up it follows that x_{126} must be 1.

Note that with these new lower and upper bounds on the variables it might pay to recompute the row bounds L_i and U_i, which again might result in tighter bounds on the variables.

2.2.2 Valid Inequalities and Cutting Planes

We present a very general procedure to construct a valid inequality for an integer program, called the *Chvatal-Gomory procedure*. The underlying idea goes back to to Gomory [36], it was later generalized by Chvatal [16] (see also Wolsey [77] for instance). We construct a valid inequality for the set $X := P \cap \mathbb{Z}^n$, where $P := \{x \in \mathbb{Q}_+^n : Ax \leq b\}$, A is an $m \times n$ matrix with columns $a_j \in \mathbb{Q}^n$ and entries $a_{ij} \in \mathbb{Q}$. Let $u \in \mathbb{Q}_+^m$. Then the inequality

$$\sum_{j=1}^n ua_j x_j \leq ub$$

(2.5)

is a valid inequality for P, because $u \geq 0$ and $\sum_{j=1}^n a_{ij}x_j \leq b_i$ for $i = 1, \ldots, n$. Since $x \geq 0$ in P we can round down the coefficients of the left-hand side of (2.5) and obtain that the inequality

$$\sum_{j=1}^n \lfloor ua_j \rfloor x_j \leq ub$$

(2.6)

is valid for P. Since $x \in \mathbb{Z}^n$ in X we can also round down the right-hand side in (2.6) and get

$$\sum_{j=1}^{n} \lfloor ua_j \rfloor x_j \leq \lfloor ub \rfloor \qquad (2.7)$$

as a valid inequality for X. Surprisingly all valid inequalities for an integer program can be generated by applying this procedure (for a proof see Chvatal [16] and Schrijver [64], or Wolsey [77]).

A valid inequality $\pi x \leq \pi_0$ for (2.1) is called *cutting plane* (or *cut*) if $\pi x \leq \pi_0$ for all feasible solutions of (2.1) and there exists an $x \in \mathbb{Q}^n$ with $Ax \leq b, \underline{x} \leq x \leq \overline{x}$ and $\pi x > \pi_0$.

Cutting planes are used to strengthen the LP relaxation of a mixed-integer program. In principle MIPs can be solved to optimality using the following cutting plane algorithm. In practice, adding too many cutting planes leads to numerical problems. Thus a stopping criterion is used to terminate the generation of cuts, even if a feasible solution is not reached.

cutAndSolve$(c, A, b, \underline{x}, \overline{x}, p)$

Input:	mixed-integer program (2.1)
(1)	**Repeat**
(2)	Solve LP relaxation of (2.1)
(3)	**If** simplex returns "feasible" **Then**
(4)	Let x^* be an optimal solution
(5)	**If** $x^* \in \mathbb{Z}^p \times \mathbb{Q}^{n-p}$ **Then**
(6)	Return x^*
(7)	**Else**
(8)	Find cutting plane (π, π_0)
(9)	Add cut to (2.1)
(10)	**End If**
(11)	**End If**
(12)	**Until** stopping criterion reached
Output:	optimal solution x^*,
	message "stopping criterion reached", or message "problem infeasible"

The first successful application of a cutting plane algorithm is due to Dantzig, Fulkerson, and Johnson [23] who solved a (at that time) large scale instance of the traveling salesman problem (see also Section 3.1).

2.2.3 Branch-and-Bound and Branch-and-Cut

Branch-and-bound algorithms for mixed integer programming use a "divide and conquer" strategy to explore the set of all feasible mixed integer solutions. But instead of exploring the whole feasible set, they make use of lower and upper bounds and therefore avoid surveying certain (large) parts of the space of feasible solutions. According to Wolsey [77] the first paper presenting a branch-and-bound strategy for the solution of integer programs is due to Land and Doig [50].

Let P_0 be a mixed-integer programming problem of form (2.1). Let $X_0 := \{x \in \mathbb{Z}^p \times \mathbb{Q}^{n-p} : Ax \leq b, \underline{x} \leq x \leq \overline{x}\}$ be the set of feasible mixed integer solutions of problem P_0. If it is too difficult to compute

$$z_{\text{MIP}} = \min_{} \quad c^T x$$
$$\text{subject to} \quad x \in X_0, \tag{2.8}$$

(for example by exploiting the combinatorial structure of X_0 as in Dijkstra's shortest path algorithm) we can split X_0 into a finite number of disjoint subsets $X_1, \ldots, X_k \subset X$, i.e., $\cup_{j=1}^k X_j = X_0$ and $X_i \cap X_j = \emptyset$ for all $i, j \in \{1, \ldots, k\}$ with $i \neq j$. Then we try to solve separately each of the subproblems

$$\min \quad c^T x$$
$$\text{subject to} \quad x \in X_j, \quad \forall\, j = 1, \ldots, k. \tag{2.9}$$

Afterwards we compare the optimal solutions of the subproblems and choose the best one. Each subproblem might be as difficult as the original problem, so one tends to solve them by the same method, i.e., splitting the subproblems again into further sub-subproblems. The (fast-growing) list of all subproblems is usually organized as a tree, called a *branch-and-bound tree*. Since this tree of subproblems looks like a family tree, one usually says that a *father* or *parent* problem is split into two or more *son* or *child* problems. This is the branching part of the branch-and-bound method.

For the bounding part of this method we assume that we can efficiently compute a lower bound $b(P)$ of some subproblem P (with feasibility set X), i.e., $b(P) \leq \min_{x \in X} c^T x$. In the case of mixed integer programming this lower bound can be obtained by using the so-called *LP relaxation* where we drop the integrality condition in (2.1) to obtain an LP, which can efficiently be solved to optimality with Dantzig's simplex algorithm. (For the ease of explanation we assume in the sequel that the LP relaxation has a finite optimum.) It occasionally happens in the course of the branch-and-bound algorithm that the optimal solution x^* of the LP relaxation of a subproblem P is also a feasible mixed integer point, i.e., it lies in X. This allows us to maintain an upper bound $U := c^T x^*$ on the optimal solution value z_{MIP} of P_0, as $z_{\text{MIP}} \leq U$. Having a good upper bound U is crucial in a branch-and-bound algorithm, because it keeps the branching tree small: Suppose the solution of the LP relaxation of some other subproblem P' satisfies $b(P') \geq U$. Then subproblem P' and further sub-subproblems derived from P' need not be considered further, because the optimal solution of this subproblem cannot be better than the best feasible solution x^* corresponding to U. The following algorithm summarizes the whole branch-and-bound procedure for mixed-integer programs.

Each (sub)problem in the list L corresponds to a node in the branch-and-bound tree, where the unsolved problems are the leaves of the tree. The node that corresponds to the entire problem (2.1) is the root node.

In the general outline of the above branch-and-bound algorithm there are two steps in the branch-and-bound part that leave some choices. In step (4) of algorithm branchAndBound we have to select a problem (node) from the list of unsolved problems to work on next, and in step (16) we must decide on how to split the problem into subproblems. A proper choice also keeps the branch-and-bound tree small and thus reduces the computation time significantly (for the details see Fügenschuh and Martin [32], for instance).

As crucial as finding a good upper bound is to find a good lower bound. Sometimes the LP relaxation turns out to be weak, but can be strengthened by adding cutting planes as discussed in Section 2.2.2. This combination of finding cutting planes and branch-and-bound leads to a hybrid algorithm called a *branch-and-cut algorithm*. In such branch-and-cut algorithm step (5) of algorithm branchAndBound is replaced by algorithm cutAndSolve.

branchAndBound($c, A, b, \underline{x}, \overline{x}, p$)

Input:	mixed-integer program $P_0 := (2.1)$
(1)	Let $L := \{P_0\}$
(2)	Let $U := +\infty$
(3)	**Repeat**
(4)	Select and remove problem P from L
(5)	Solve LP relaxation of P
(6)	**If** simplex returns "feasible" **Then**
(7)	Let x^* be an optimal solution
(8)	Let $b(P) := c^T x^*$
(9)	**If** $x^* \in \mathbb{Z}^p \times \mathbb{Q}^{n-p}$ **Then**
(10)	**If** $U > b(P)$ **Then**
(11)	Let $U := b(P)$
(12)	$\tilde{x} := x^*$
(13)	Delete from L all subproblems P with $b(P) \geq U$
(14)	**End If**
(15)	**Else**
(16)	Split problem P into subproblems
(17)	Add subproblems to L
(18)	**End If**
(19)	**End If**
(20)	**Until** $L = \emptyset$
Output:	optimal solution $\tilde{x}$ for (2.1) or message "problem infeasible" (if $U = \infty$)

For more informations about branch-and-bound and branch-and-cut see also Nemhauser and Wolsey [59] or Fügenschuh and Martin [32], for instance.

2.2.4 Multicriteria Mixed-Integer Programs

In this section we take a look into the multicriteria optimization tool-box and introduce some basic definitions and techniques. For further details we refer to the book of Ehrgott [29], for instance.

The mixed-integer program (2.1) has a unique objective function $f := f_1$. When dealing with real-world decisions we are often faced with the problem that several goals (i.e., objective functions) $f := (f_1, \ldots, f_q)$ have to be minimized at the same time. In many cases these goals are in conflict with each other.

If $X := \{x \in \mathbb{Z}^p \times \mathbb{Q}^{n-p} : Ax \leq b, \underline{x} \leq x \leq \overline{x}\}$ denotes the set of feasible solutions for some MIP then the set of *outcomes* is defined as $Y := f(X)$. A feasible solution $x \in X$ is called *dominated* by $x^* \in X$ if $f(x^*) < f(x)$, i.e., $f_i(x^*) \leq f_i(x)$ for all $i = 1, \ldots, q$ and $f(x^*) \neq f(x)$. A feasible solution x^* is *Pareto optimal*, if there is no $x \in X$ that dominates x^*. Denote X_{Par} the set of Pareto optimal solutions of X with respect to f, then $Y_{\text{eff}} = f(X_{\text{Par}})$ is called the *efficient set*. All other outcomes in $Y \backslash Y_{\text{eff}}$ are called *non-efficient*. Having these definitions at hand, we can state a multicriteria optimization problem as:

$$\begin{aligned} \min \quad & f(x) \\ \text{subject to} \quad & x \in X. \end{aligned} \tag{2.10}$$

Solving problem (2.10) means the computation of Pareto optimal solutions. Several different techniques for this are discussed in Ehrgott's book [29]. In the following, we present two of the most common ones: Lexicographic optimality and weighted sum scalarization.

A solution $x^* \in X$ is called *lexicographically optimal*, if $f(x^*)$ is a lexicographically minimal vector in Y, i.e., $f_i(x^*) < f_i(x)$ for all $x \in X$ and $i = \min\{j : f_j(x^*) \neq f_j(x)\}$. It can be shown that a lexicographically optimal solution is also Pareto optimal. This concept is applied, if there exists an ordering among the objectives such that objective f_i is much (infinitely) more important than f_{i+1} for all $i = 1, \ldots, q-1$. The other concept is the *weighted sum scalarization*, where the objectives of (2.10) are replaced by a single linear objective:

$$\begin{aligned} \min \quad & \sum_{i=1}^{q} \lambda_i f_i(x) \\ \text{subject to} \quad & x \in X, \end{aligned} \tag{2.11}$$

with

$$\lambda_1, \ldots, \lambda_q > 0, \quad \lambda_1 + \ldots + \lambda_q = 1. \tag{2.12}$$

In contrast to the lexicographical optimality, weighted sum scalarization is used if there is no obvious ordering among the objectives, and a gluing of different goals into a single objective is justifiable. Problem (2.11) is a classical single-objective mixed-integer program (MIP). It can also be shown that every optimal solution x of (2.11) for any λ fulfilling (2.12) is a Pareto optimal solution. However, the converse is not true in general. Note that lexicographically optimal solutions can theoretically be computed by weighted sum scalarization if X is a bounded set and each coefficient λ_i is sufficiently greater than λ_{i+1} for all $i = 1, \ldots, q - 1$.

Having a Pareto optimal solution at hand, it is natural to ask how much a certain single objective can at most be improved, and conversely, how much can it be deteriorated? The two reference points $\bar{\varepsilon} = (\bar{\varepsilon}_1, \ldots, \bar{\varepsilon}_q)$ given by

$$\bar{\varepsilon}_i := \begin{array}{c} \max \quad f_i(x) \\ \text{subject to} \quad x \in X \end{array} \tag{2.13}$$

for $i = 1, \ldots, q$, and $\underline{\varepsilon} = (\underline{\varepsilon}_1, \ldots, \underline{\varepsilon}_q)$ given by

$$\underline{\varepsilon}_i := \begin{array}{c} \min \quad f_i(x) \\ \text{subject to} \quad x \in X, \end{array} \tag{2.14}$$

for $i = 1, \ldots, q$, yield valuable information for a comparison. In the literature the point $\underline{\varepsilon}$ is also called *ideal point* (the point $\bar{\varepsilon}$ does not have a particular name). Its coefficients are a tight bound on the set of Pareto outcomes.

2.2.5   Integer Programs with Two Variables per Row

In the past two and half decades several researchers devoted their studies to linear inequality systems where each constraint has at most two non-zero coefficients. As it will turn out in the sequel an important substructure for the models presented in this thesis are exactly these inequality systems. Moreover, in these inequalities one coefficient is positive and the other is negative. In this section, we review some relevant articles from this area, with a special focus on those results and techniques we will frequently use in the sequel.

Consider an integer program of the following type:

$$\text{(IP2)} \quad \begin{array}{l} \min \quad w^T x \\ \text{subject to} \quad a_k x_{i_k} + b_k x_{j_k} \leq c_k, \quad \forall\, k \in \{1, \ldots, m\}, \\ \quad \underline{x} \leq x \leq \bar{x} \\ \quad x \in \mathbb{Z}^n. \end{array} \tag{2.15}$$

with $1 \leq i_k, j_k \leq n$, $w \in \mathbb{Q}^n$, $a, b, c \in \mathbb{Q}^m$, $\underline{x}, \bar{x} \in \mathbb{Z}^n$. This problem is NP-hard, because it is a generalization of the well-known minimum weight vertex cover problem (VC) and the minimum weight 2-satisfiability problem (2SAT), which are both known to be NP-hard, see Garey and Johnson [34].

System (IP2) is called *monotone* if $a_k \cdot b_k < 0$ for all $k \in \{1, \ldots, m\}$, i.e., in each inequality one coefficient is positive and the other is negative. From a result of Lagarias [49] on the NP-completeness of the simultaneous Diophantine approximation problem it follows that finding a feasible solution of a monotone system is also NP-complete. An instance of this problem consists of a rational vector $\alpha = (\frac{a_1}{b_1}, \ldots, \frac{a_n}{b_n})$, and positive integers N, s_1, and s_2. The question is whether there exists an integer Q with $1 \leq Q \leq N$, such that

$$\max\left\{\delta\left(\frac{Qa_i}{b_i}\right) : 1 \leq i \leq n\right\} \leq \frac{s_1}{s_2}, \tag{2.16}$$

where $\delta(q)$ denotes the distance of q to the nearest integer, i.e., $\delta(q) := \min\{q - n : n \in \mathbb{Z}\}$. This problem can be formulated as an instance of finding an integer feasible solution $(x_1, \ldots, x_n, Q)$ for the following monotone system:

$$1 \leq Q \leq N \tag{2.17}$$

$$-s_1 \leq s_2\left(\frac{a_i}{b_i}Q - x_i\right) \leq s_1, \quad \forall\, 1 \leq i \leq n. \tag{2.18}$$

If we drop the integrality condition in (IP2), then we obtain (LP2), the linear programming relaxation of (IP2) with two variables per constraint. The problem of computing a feasible solution in this case has also been investigated intensively. It was observed by Pratt [62] that the feasibility of both (LP2) and (IP2) can be decided quickly by examining the cycles in certain graphs, if one coefficient is 1 and the other is -1 (i.e., the dual of a shortest-path problem). For this special case he gave an algorithm with $O(n^3)$ time complexity. Shostak [68] generalized Pratt's method to feasibility of an inequality system (LP2) with two variables and arbitrary coefficients, and further to feasibility of arbitrary sets of linear inequalities. However, Shostak's method has an exponential worst-case behavior. In both Pratt's and Shostak's method, a vertex is introduced for every variable of the linear program, together with an additional vertex x_0. Every inequality with two non-zero coefficients is represented by an edge between the corresponding pair of vertices. Inequalities involving only one variable (i.e., lower and upper bounds on the variables) are represented by an edge to and from vertex x_0. The graph then consists of $n + 1$ vertices and m edges; there may be multiple edges between some vertices. Polynomial-time algorithms for the feasibility of (LP2) were given by Aspvall and Shiloach [5]. Their algorithm requires $O(mn^3I)$ arithmetic operations, where I is the input length of the problem data. Megiddo [55] presented the first full-polynomial algorithm (i.e., an algorithm not depending on I) for the feasibility of (LP2) which requires $O(mn^3\log m)$ operations. A better algorithm for this problem is due to Cohen and Megiddo [19], with a running time of $O(mn^2(\log n + \log m))$. The main common feature of all these algorithms is the determination of upper and lower bounds on each variable by following paths and cycles in graphs. A completely different technique was introduced by Nelson [58] who gave an $O(mn^{\lceil \log_2 n\rceil + 4}\log n)$ algorithm to check feasibility for (LP2). His algorithm uses a special tailored Fourier-Motzkin elimination method as backbone. The fastest algorithm for (LP2) so far was found by Hochbaum and Naor [41], with running time of $O(mn^2\log m)$. It makes use of some ideas of Aspvall and Shiloach as well as Fourier-Motzkin elimination steps.

Since (IP2) is NP-complete, no polynomial time algorithm for finding feasible solutions is expected, unless $P = NP$. For monotone inequality systems, Hochbaum et al. [40] gave a pseudo-polynomial 2-approximative algorithm that runs in $O(mnU^2\log(Un^2m))$, where $U := \max\{\overline{x}_i - \underline{x}_i : i = 1, \ldots, n\}$. In [41] Hochbaum and Naor extended this algorithm to the case of non-monotone systems. This algorithm is also 2-approximative and has the same running time behavior. It is based on their observation that a feasible solution for (IP2) can be identified in $O((n + m)U)$ time. Bar-Yehuda and Rawitz [7] recently improved these results and gave simple feasibility and optimality algorithms for monotone and non-monotone (IP2) systems. The next two routines from their article turned out to be useful for our application.

Consider the constraint $a_k x_{i_k} + b_k x_{j_k} \leq c_k$ of (IP2). We have to distinguish four different cases concerning the signs of a_k and b_k. For example, if $a_k > 0$ and $b_k > 0$ then we obtain

$$b_k x_{j_k} \leq c_k - a_k x_{i_k} \leq c_k - a_k \underline{x}_{i_k}. \tag{2.19}$$

Dividing by b_k yields

$$x_{j_k} \leq \frac{c_k - a_k \underline{x}_{i_k}}{b_k}. \tag{2.20}$$

Since all x variables are integer we can round down the right hand side to obtain a possibly better bound:

$$x_{j_k} \leq \lfloor \frac{c_k - a_k \underline{x}_{i_k}}{b_k} \rfloor. \tag{2.21}$$

Now we compare this bound with the previous upper bound $\overline{x}_{j_k}$ on x_{j_k} and select the lower one as new upper bound on this variable. That is, we set

$$\overline{x}_{j_k} := \min\{\overline{x}_{j_k}, \lfloor \frac{c_k - a_k \underline{x}_{i_k}}{b_k} \rfloor\}. \tag{2.22}$$

The other three cases are similar to this one. Hence we omit the details of their derivations.

The impact of this constraint and the bounds $\underline{x}_{i_k}, \overline{x}_{i_k}$ on the bounds $\underline{x}_{j_k}, \overline{x}_{j_k}$ (and vice versa) is evaluated by the routine oneOnOneImpact.

oneOnOneImpact$(\underline{x}, \overline{x}, i, j, k)$	
Input:	bounds $\underline{x}, \overline{x}$, variable indices i, j, inequality index k
(1)	If $a_k > 0$, **Then**
(2)	If $b_k > 0$, **Then**
(3)	$\overline{x}_{j_k} := \min\{\overline{x}_{j_k}, \lfloor \frac{c_k - a_k \underline{x}_{i_k}}{b_k} \rfloor\}$
(4)	**Else**
(5)	$\underline{x}_{j_k} := \max\{\underline{x}_{j_k}, \lceil \frac{c_k - a_k \overline{x}_{i_k}}{b_k} \rceil\}$
(6)	**End If**
(7)	**Else**
(8)	If $b_k > 0$, **Then**
(9)	$\overline{x}_{j_k} := \min\{\overline{x}_{j_k}, \lfloor \frac{c_k - a_k \overline{x}_{i_k}}{b_k} \rfloor\}$
(10)	**Else**
(11)	$\underline{x}_{j_k} := \max\{\underline{x}_{j_k}, \lceil \frac{c_k - a_k \underline{x}_{i_k}}{b_k} \rceil\}$
(12)	**End If**
(13)	**End If**
Output:	updated bounds $\underline{x}, \overline{x}$

Note that oneOnOneImpact is a special case of bounds strengthening (the general MIP preprocessing technique presented in Section 2.2.1) for IP2.

The next routine oneOnAllImpact evaluates the impact of $\underline{x}_t, \overline{x}_t$ on all other bounds and changes $\underline{x}, \overline{x}$ accordingly.

oneOnAllImpact($\underline{x}, \overline{x}, t$)
Input: bounds $\underline{x}, \overline{x}$, variable index t
(1)　Let $L := \{t\}$
(2)　**While** $L \neq \emptyset$ **Do**
(3)　　　Select and remove i from L
(4)　　　**For** each constraint k with x_i and another variable x_j **Do**
(5)　　　　　Call OneOnOneImpact($\underline{x}, \overline{x}, i, j, k$)
(6)　　　　　**If** $\underline{x}_j > \overline{x}_j$ **Then**
(7)　　　　　　　Return "system infeasible"
(8)　　　　　**End If**
(9)　　　　　**If** $\underline{x}_j$ or $\overline{x}_j$ changed **Then**
(10)　　　　　　Let $L := L \cup \{j\}$
(11)　　　　　**End If**
(12)　　　**End Do**
(13)　**End Do**
Output: improved bounds $\underline{x}, \overline{x}$ or message "system infeasible"

Using these two routines Bar-Yehuda and Rawitz achieved the following theorem (for a proof see [7]).

Theorem 1 *Given an (IP2) with m inequalities, n variables and $U := \max\{\overline{x}_i - \underline{x}_i :$ $i = 1, \ldots, n\}$. If (IP2) is monotone then either an optimal solution can be computed, or it can be shown that no solution exists (both in $O(mU)$ time). If (IP2) is non-monotone then either a feasible solution can be computed, or it can be shown that no solution exists (both in $O(mU)$ time).*

2.2.6 Irreducible Infeasible Subsystems

Let $X := \{x \in \mathbb{Z}^p \times \mathbb{Q}^{n-p} : Ax \leq b, \underline{x} \leq x \leq \overline{x}\}$ be the set of feasible solutions for some MIP. If $X = \emptyset$ then some of the constraints might contradict each other. In this case it is natural to ask which of the inequalities describing X actually don't fit to each other. A subsystem of the inequalities that is infeasible but becomes feasible if any of the inequalities is removed is called *irreducible infeasible system* or *irreducible inconsistent system* (IIS, for short). The detection of IIS was studied by several authors before, mainly in the context of linear programming problems, see for instance the work of van Loon [74], Chinneck and Dravnieks [15], or Tamiz, Mardle, and Jones [70]. In the case of linear inequalities having integer variables, which is computationally more demanding, the problem of finding an IIS has been approached by means of heuristics, see Guieu and Chinneck [38] or Bruni [14].

In the sequel, we present an algorithm for the selection of an IIS for general inequality systems. Our algorithm is based on sorting the inequalities in an appropriate way, and an oracle to determine feasibility of a given set of inequalities. We finally apply this general algorithm to integer programs with at most two non-zero coefficients per constraint (IP2).

Given an infeasible linear inequality system $Ax \leq b$ (which now includes the trivial inequalities $\underline{x} \leq x \leq \bar{x}$ as well as the integrality constraints), we use the list $L := (i_1, \ldots, i_m)$ as an abbreviation for its inequalities. In the j-th step of the algorithm, infeasibility is checked (by calling the oracle) for the subsystems consisting of the first k_j inequalities, for every $k_j = j, \ldots, k_{j-1}$ (with $k_0 := m$). For some k_j infeasibility will be detected. Then we shift every inequality from j to $k_j - 1$ one position to the right within the list L, and set inequality k_j on the free position j in the inequality system. We then check feasibility of the first j inequalities. If this check returns "infeasible" then an IIS is found: $(i_{k_1}, \ldots, i_{k_j})$. If not, we increase j by 1 and repeat this procedure, until an IIS is found.

findIIS$(i_1, \ldots, i_m)$

Input:	inequality system $L := (i_1, \ldots, i_m)$
(1)	Let $j := 1, k_0 := m$
(2)	**For** k_j **From** j **To** k_{j-1} **Do**
(3)	Call oracle to check the first k_j inequalities
(4)	**If** oracle returns "infeasible" **Then**
(5)	**If** $k_j > j$ **Then**
(6)	Shift every inequality from j to $k_j - 1$ to the right in list L
(7)	Set inequality k_j on the (free) j-th position
(8)	**End If**
(9)	**If** $j > 1$ **Then**
(10)	Call oracle to check the first j inequalities
(11)	**If** oracle returns "infeasible" **Then**
(12)	Return IIS : $i_{k_1}, \ldots, i_{k_j}$
(13)	**End If**
(14)	**End If**
(15)	Let $j := j + 1$
(16)	Go to step (2)
(17)	**End If**
(18)	**End For**
(19)	Return "system feasible"
Output:	IIS $i_{k_1}, \ldots, i_{k_j}$ or message "system feasible"

It remains to show the correctness of the algorithm, i.e., we have to show that the returned subsystem of inequalities is indeed an IIS.

Theorem 2 *The output of algorithm* findIIS *is correct.*

Proof. Suppose the algorithm returns the system $(i_{k_1}, \ldots, i_{k_j})$ for some j. We have to show that for every j' with $1 \leq j' \leq j$ the system $(i_{k_1}, \ldots, i_{k_{j'-1}}, i_{k_{j'+1}}, \ldots, i_{k_j})$ is feasible. Consider the j'-th step of the algorithm. In this step, the entire system had the following order: $(i_{k_1}, \ldots, i_{k_{j'-1}}, i_{a_1}, \ldots, i_{a_{l'}}, i_{k_{j'}}, i_{b_1}, \ldots, i_{b_{l''}})$, with $a_1 \leq \ldots \leq a_{l'} \leq k_{j'} \leq b_1 \leq \ldots \leq b_{l''}$, and infeasibility was first detected for the subsystem: $(i_{k_1}, \ldots, i_{k_{j'-1}}, i_{a_1}, \ldots, i_{a_{l'}}, i_{k_{j'}})$. Remark that $k_1 \geq k_2 \geq \ldots \geq k_{j'-1} \geq k_{j'} \geq k_{j'+1} \geq \ldots \geq k_j$ and, by the definition of the algorithm, $k_{j'+1}, \ldots, k_j \in \{a_1, \ldots, a_{l'}\}$. Thus, $(i_{k_1}, \ldots, i_{k_{j'-1}}, i_{a_1}, \ldots, i_{a_{l'}})$ is a feasible subsystem, which itself contains $(i_{k_1}, \ldots, i_{k_{j'-1}}, i_{k_{j'+1}}, \ldots, i_{k_j})$ as a (feasible) subsystem. $\square$

We give a small example to demonstrate how the findIIS algorithm works. Consider the infeasible inequality system

$$
\begin{array}{rrrcrcr}
i_1 : & x_1 & & & & \geq & 0 \\
i_2 : & -x_1 & - & x_2 & & \geq & -6 \\
i_3 : & -3x_1 & - & 2x_2 & & \geq & -6 \\
i_4 : & -x_1 & + & 2x_2 & & \geq & 2 \\
i_5 : & & - & x_2 & & \geq & -2 \\
i_6 : & x_1 & - & x_2 & & \geq & 0
\end{array}
\tag{2.23}
$$

Since this inequality system is an integer program with two non-zero coefficients per

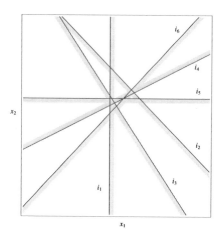

Figure 2.1: A plot of inequality system (2.23).

constraint we can use the Bar-Yehuda and Rawitz algorithm (Theorem 1) as an oracle to determine the feasibility of subsystems.

In the first step, feasibility is checked for the first k_1 inequalities of the system $(i_1, i_2, i_3, i_4, i_5, i_6)$, for $k_1 = 1, \ldots, 6$. Infeasibility first occurs for $k_1 = 6$, so the new order of the inequality system is $(i_6, i_1, i_2, i_3, i_4, i_5)$. In step two, feasibility is checked for the first k_2 inequalities, for $k_2 = 2, \ldots, 6$. Infeasibility first occurs for $k_2 = 5$, so the new order of the system is $(i_6, i_4, i_1, i_2, i_3, i_5)$. Since the subsystem (i_6, i_4) is feasible, the algorithm continues. In step three, feasibility is checked for the first k_3 inequalities, for $k_3 = 3, \ldots, 5$. Infeasibility occurs only for $k_3 = 5$, so we obtain $(i_6, i_4, i_3, i_1, i_2, i_5)$. After a final feasibility check of the subsystem (i_6, i_4, i_3) the algorithm returns with this subsystem as an IIS.

We count the number of calls of the feasibility oracle in the worst case. In this case, the IIS consists of all inequalities of the system, and their infeasibility is always detected at the last inequality. Thus in step $k \leq m$ of the algorithm we need $m - k$ calls of the oracle, therefore a total number of $\frac{m(m-1)}{2}$ calls.

Combining this with Theorem 1 we have just shown the following results:

Theorem 3 *If an infeasible (IP2) is given, then an IIS of (IP2) can be found in $O(m^3 U)$ time.*

This worst-case bound is sharp, as the following example shows. Consider the inequality system

$$
\begin{array}{llrcccl}
i_1 & : & x_1 & - & x_2 & \geq & 0 \\
i_2 & : & x_2 & - & x_3 & \geq & 0 \\
\vdots & & & & & & \vdots \\
i_{n-1} & : & x_{n-1} & - & x_n & \geq & 0 \\
i_n & : & x_n & - & x_1 & \geq & 1
\end{array}
\tag{2.24}
$$

It is easy to see that this system is infeasible, because adding all inequalities yields $0 \geq 1$. Moreover, the entire system is an IIS. However, the findIIS algorithm is only convinced of this fact after having permuted the inequality system to $(i_n, \ldots, i_2, i_1)$.

Chapter 3

Vehicle Routing with Coupled Time Windows

The integrated planning problem IOSANA and our models for IOSANA presented in Chapter 4 belong to the large family of vehicle routing and scheduling problems, starting from the elementary traveling salesman problem (TSP) in Section 3.1 to the case of multiple traveling salesmen, the vehicle routing problem (VRP), in Section 3.2. For many applications, vehicle routing models have to be extended to include aspects that are essential to the routing of vehicles in a real-life world. The family of those extended problems is called Rich Vehicle Routing Problems (RVRP). In this sense we extend the vehicle routing problem to what we call the vehicle routing problem with coupled time windows (VRPCTW) in Section 3.3. This model will serve as a basis for IOSANA. We conclude this chapter with some remarks on the theoretical complexity of the VRPCTW in Section 3.4.

3.1 The Traveling Salesman Problem

The simplest and perhaps the most famous member of the vehicle routing family is the classical and well-known *traveling salesman problem*, or *TSP*, for short. As the name indicates, the TSP is a model for real-world problems where a number of customers must be visited by a single salesman and the traveling time (or traveling costs) between two customers is the crucial factor. An instance of the TSP consists of a graph $G = (V, E)$ with vertex set $V = \{1, \ldots, n\}$ and edge set E, and non-negative weights $c_{vw} \in \mathbb{Q}_+$ on each edge $\{v, w\} \in E$. The problem asks for a minimum-weight Hamilton circuit (called *tour*).

Historically it is not exactly known when mathematicians got interested in studying the TSP. The oldest problem formulation (without mathematics) is perhaps a manual for the successful traveling salesman from 1832 (found by Müller-Merbach [57], see also Schrijver [65]). According to Schrijver [65] the first mathematician who has written about the TSP was Menger in 1928. It was also Menger who brought this problem to mathematicians' attention in the USA in the early 1930s, when it was finally picked up by Dantzig, Fulkerson and Johnson, working together at RAND at that time. In their seminal paper [23] they applied for the first time Dantzig's at that time new simplex method to solve a difficult (*NP*-hard, as we say today) optimization problem, where they in particular introduced the cutting plane technique, and solved an instance with 42 vertices (see Figure 3.1). (The reader who is interested in further details of this story is referred to the excellent historical report of Schrijver [65] on the development of combinatorial optimization till 1960.)

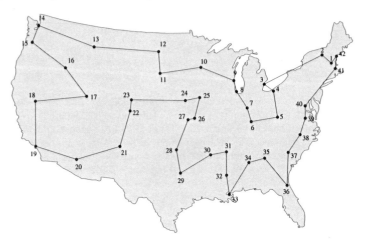

Figure 3.1: The 42-cities instance of [23] and its optimal solution.

The decision variant of the TSP, i.e., the question whether there is a solution of weight less or equal a given value, is *NP*-complete, for it can be transformed from the Hamiltonian circuit problem, see Garey and Johnson [34] for the details. Despite its theoretical difficulty, instances of the traveling salesman problem with up to several thousand vertices can today be solved to optimality using good primal heuristics for solution finding and an LP-based branch-and-cut approach to prove optimality (see Applegate et al. [2] for a survey and www.tsp.gatech.edu for the most recent results).

When the underlying graph is a digraph we speak of the asymmetric traveling salesman problem (ATSP). An instance of the ATSP consists of a graph $G = (V, A)$ with vertex set $V = \{1, \ldots, n\}$ and arc set A, and non-negative weights $c_{vw} \in \mathbb{Q}_+$ on each arc $(v, w) \in A$. As in the TSP, also the ATSP problem asks for a minimum-weight Hamilton circuit (also called *tour*). This problem can be formulated as a linear integer program with binary variables. For every arc $(v, w) \in A$ introduce a binary variable $x_{vw} \in \{0, 1\}$, with $x_{vw} = 1$ if (v, w) is an arc of the Hamiltonian circle. The requirement that each vertex occurs exactly once in the tour is formulated as

$$\sum_{v:(v,w)\in A} x_{vw} = 1, \tag{3.1}$$

for all $w \in V$, that is, every vertex $w \in V$ is "entered" exactly once, and

$$\sum_{w:(v,w)\in A} x_{vw} = 1, \tag{3.2}$$

for all $v \in V$, that is, every vertex $v \in V$ is "left" exactly once. The constraints (3.1) and (3.2) are not sufficient to define tours within an integer model of the ATSP. Consider for example the instance with $n = 6$ shown in Figure 3.2, where $x_{15} = x_{56} = x_{61} = x_{23} = x_{34} = x_{42} = 1$ is a feasible solution for inequalities (3.1) and (3.2) (every vertex is entered and left exactly once), but is not a tour. One way to eliminate such solutions

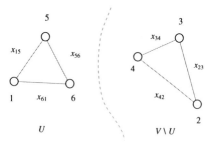

Figure 3.2: Two subcycles and the corresponding cut.

is to introduce an additional constraint saying there must be an arc connecting $\{1, 5, 6\}$ with $\{2, 3, 4\}$. In general these constraints are:

$$\sum_{(v,w)\in A, v\in U, w\in V\setminus U} x_{vw} \geq 1, \tag{3.3}$$

for all $U \subset V$ with $2 \leq |U| \leq |V| - 2$. Inequalities (3.3) are satisfied by all tours, but any subtour violates at least one of them.

The objective is to find a Hamiltonian circuit with the least weight. That is, we want to minimize

$$\sum_{(v,w)\in A} c_{vw} x_{vw}. \tag{3.4}$$

Thus a model for the ATSP can be stated as follows:

$$\begin{aligned} \min \quad & (3.4) \\ \text{subject to} \quad & (3.1), (3.2), (3.3) \\ & x \in \{0, 1\}^{|A|}. \end{aligned} \tag{3.5}$$

There are $2^{|V|}$ subtour elimination constraints, just too many to be explicitly written down. From a practical point of view this in not a major drawback, if formulation (3.5) is solved within a branch-and-cut approach. In this case only a few of the subtour elimination constraints are generated on demand as cutting planes during the search of the optimal solution.

For the ATSP there are other formulations known which need a much smaller number of inequalities. One of these formulations was proposed by Miller, Tucker, and Zemlin [56] (see also Nemhauser and Wolsey [59]): We introduce an additional variable $t_v \in \mathbb{Q}_+$, for each vertex $v \in V \backslash \{1\}$, and consider the constraints

$$t_v + 1 - n \cdot (1 - x_{vw}) \le t_w \tag{3.6}$$

for all $(v, w) \in A$ with $v, w \ne 1$. We now show that the inequalities (3.6) are also able to cut off all subtours from the set of feasible solutions. Suppose $x \in \{0, 1\}^{|A|}$ is a vector representing (at least) two subtours. Then one of the subtours contains vertex 1, the other(s) not. If we sum (3.6) over the arc set A' of those arcs corresponding to some subtour that does not contain vertex 1, then we get

$$\sum_{(v,w) \in A'} x_{vw} \le |A'| \cdot \left(1 - \frac{1}{n}\right), \tag{3.7}$$

which means that all subtours not containing vertex 1 are forbidden. Hence a feasible solution cannot have any subtour at all. Next we have to show that no tour is excluded. Suppose $x \in \{0, 1\}^{|A|}$ represents a tour. For all $v \in V$ set $t_v := p(v)$, where $p(v)$ is the position of vertex v in the tour of x. We distinguish two cases, $x_{vw} = 0$ and $x_{vw} = 1$. In the first case, $x_{vw} = 0$, we obtain $t_v - t_w + n \cdot x_{vw} = t_v - t_w = p(v) - p(w) \le n - 2$, because the largest distance between v and w in the tour is $n - 2$. Thus (3.6) holds. In the second case, $x_{vw} = 1$, w is directly after v in the tour, hence $p(w) = p(v) + 1$, and we obtain $t_v + 1 - n \cdot (1 - x_{vw}) = t_v + 1 = p(v) + 1 = p(w) = t_w$, so again (3.6) holds. Together we have shown that

$$\begin{aligned} \min \quad & (3.4) \\ \text{subject to} \quad & (3.1), (3.2), (3.6) \\ & x \in \{0, 1\}^{|A|}, t \in \mathbb{Q}_+^{|V|-1} \end{aligned} \tag{3.8}$$

is another model for the ATSP having significantly less inequalities than (3.5). On the other hand, the LP-relaxation of the latter formulation is much weaker: Consider for example an instance with $n \ge 5$, and set $t_3 = t_4 = t_5 = 0$ and $x_{34} = x_{45} = x_{53} = \frac{n-1}{n}$. Then (3.6) is satisfied by this solution, but not (3.3): If we take $U = \{1, 2\}$, then from $\frac{n-1}{n} > \frac{2}{3}$ we get $\sum_{(v,w) \in A, v \in U, w \in V \backslash U} x_{vw} < 1$.

However, there is an advantage in the use of model (3.8), because it can be easily extended to the case of time windows. That is, in some application every customer (vertex) may specify an earliest and latest arrival time of the salesman. We replace the value 1 in inequality (3.6) by some parameter $d_{vw} \in \mathbb{Q}_+$ for every arc $(v, w) \in A$ with $v, w \ne 1$, then we obtain

$$t_v + d_{vw} - M \cdot (1 - x_{vw}) \le t_w, \tag{3.9}$$

where M is a sufficiently large value (for example, set $M := \max_{v \in V}\{\bar{t}_v\} + \max_{(v,w) \in A}\{d_{vw}\}$). The parameter d_{vw} represents the time the salesman stays at customer v (to fulfill some duty) plus the time he needs to travel from v to w. For all vertices $v \in V \backslash \{1\}$ we add lower and upper bounds on variable t_v:

$$\underline{t}_v \le t_v \le \bar{t}_v. \tag{3.10}$$

Then the mixed-integer program

$$
\begin{array}{ll}
\min & (3.4) \\
\text{subject to} & (3.1), (3.2), (3.9) \\
& (3.10) \\
& x \in \{0,1\}^{|A|}, t \in \mathbb{Q}_+^{|V|-1}
\end{array}
\tag{3.11}
$$

is a model for the so-called *traveling salesman problem with time windows* (TSPTW). The TSPTW is also an NP-hard problem, because it is a generalization of the ATSP.

On the other hand, the ATSP can be seen as a relaxation for the TSPTW, because no time windows at the vertices have to be respected. Hence it may happen that an instance of the TSPTW has no solution, but the corresponding ATSP relaxation has one. This indicates that a single salesman just is not enough to travel to all customers at the right time, and this motivates further generalizations of the TSPTW to the case of multiple salesmen.

3.2 The Vehicle Routing Problem

In many real-world applications a single traveling salesman might not be able to handle all customers due to time or capacity restrictions. As a mathematical problem, the vehicle routing problem (VRP) was introduced in 1959 by Dantzig and Ramser [24]. They described a real-world application of delivering gasoline to service stations, proposed a mathematical programming formulation for the problem, and gave a heuristic solution algorithm. Later, in 1964, Clarke and Wright [18] improved the Dantzig-Ramser heuristic and gave an effective greedy heuristic, which is (according to Laporte and Semet [51]) "perhaps the most widely known heuristic for the VRP". These two seminal papers were followed by several hundreds of models and algorithms to solve or approximate solutions for various different versions of the VRP. On one hand this research is motivated by the great practical relevance of the VRP, and on the other hand by its difficulty. Both the VRP and the TSP (and also the ATSP) are NP-hard combinatorial problems, thus in some sense they are "equally difficult". However, from a practical point-of-view it is possible to solve TSP instances with some ten thousand vertices, whereas the most effective exact algorithms for the VRP today might fail in solving problems with more than 70 vertices.

Among the large amount of different VRP models and variants it is difficult to find a somehow "generic VRP" that is representative for the whole class. Depending on the particular application there are basically two reasons why a single vehicle might not be sufficient. Either there is a load to be picked up at the customers' sites and the load capacity of the vehicle is crucial, or there are time windows in which the arrival of the vehicle must happen to satisfy the customers' needs. Toth and Vigo [71] establish the first problem, also called *capacitated vehicle routing problem* (CVRP) as the most generic VRP of them all. However, for the problem presented in this thesis, capacities do not have to be issued, but time window related aspects are crucial. Hence we consider here the *vehicle routing problem with time windows* (VRPTW) as the generic VRP.

An instance of the VRPTW is defined as follows. Let $G = (V, A)$ be a graph with node set $V = \{1, \ldots, n\}$ and arc set A. Let c_{vw} be non-negative weights on each arc $(v, w) \in A$ representing the costs (in terms of time or money) for driving from v to w. Let $1 \in V$ be the *depot node* where all vehicles start and end their trips. For every other node $v \in V \backslash \{1\}$ a time window $\underline{t}_v, \bar{t}_v$ with $\underline{t}_v \leq \bar{t}_v$ is given, in which a vehicle has to arrive. If a vehicle arrives at node v earlier then it has to wait until $\underline{t}_v$. The non-negative parameter d_{vw} stands for the time the vehicle stays at v (where the driver fulfills some duty) plus the time for driving from v to w. What we seek is either a set of optimal tours for a given number of vehicles, a minimum number of vehicles that is able to serve all customers in time, or a mixture of these two goals.

We now give a mathematical programming formulation for the VRP with time windows. The VRPTW uses the same variables as the TSPTW, namely a binary variable $x_{vw} \in \{0, 1\}$ for every arc $(v, w) \in A$, with $x_{vw} = 1$ if (v, w) occurs in some tour of a solution, and a non-negative variable $t_v \in \mathbb{Q}_+$ for every node $v \in V \backslash \{1\}$ representing the starting time of the service at vertex v.

The main difference between the TSPTW and the VRPTW is that the depot vertex $1 \in V$ is entered

$$\sum_{v:(v,1)\in A} x_{v1} = K \tag{3.12}$$

and left

$$\sum_{w:(1,w)\in A} x_{1w} = K \tag{3.13}$$

exactly K times, where $K \in \mathbb{Z}_+$ is either a parameter or a variable depending on whether the size of the fleet is fixed or not, respectively. Note that for $K = 1$ we obtain the TSPTW as a special case of the VRPTW. As in the TSPTW every other vertex $w \in V \backslash \{1\}$ is entered exactly once:

$$\sum_{v:(v,w)\in A} x_{vw} = 1, \tag{3.14}$$

and every vertex $v \in V \backslash \{1\}$ is also left exactly once:

$$\sum_{w:(v,w)\in A} x_{vw} = 1. \tag{3.15}$$

If we assume that the size of the fleet is variable, then the two objectives are:

- A reduction of the fleet-size:

$$K, \tag{3.16}$$

- and a minimal total length of all tours:

$$\sum_{(v,w)\in A} c_{vw} x_{vw}. \tag{3.17}$$

Then the VRPTW can be stated as the following bicriteria mixed-integer program:

$$
\begin{aligned}
\min \quad & ((3.16),(3.17)) \\
\text{subject to} \quad & (3.9),(3.12),\dots,(3.15) \\
& (3.10) \\
& x \in \{0,1\}^{|A|}, K \in \mathbb{Z}_+, t \in \mathbb{Q}_+^{|V|-1}.
\end{aligned} \tag{3.18}
$$

3.3 The Vehicle Routing Problem with Coupled Time Windows

Several variants and extensions of this basic VRP version (3.18) are described in the literature, depending on the particular application of the model. The family of those extended problems is called *Rich Vehicle Routing Problems* (RVRP). Interesting occurrences of this richness include, among others: The use of heterogeneous vehicles, the limitation due to load capacities, periodic vehicle routing, backhauling, pickup and delivery, the fleet size and mix problem, the use of vehicles with multiple compartments, robustness in routing solutions, or the management of multiple objectives in routing schedules.

For example, the VRP with backhauls (VRPB), see for instance Toth and Vigo [72], is an extension of the VRP where the set of customers is divided into two disjoint subsets, $V = L \cup B, L \cap B = \emptyset$. The first one, L, contains the linehaul customers, the second subset, B, contains the backhaul customers. Linehaul customers request a certain quantity of a product to be delivered, whereas at backhaul customers, certain quantities of a product can be picked up. Moreover, a precedence constraint exists: If a vehicle serves both, backhaul and linehaul customers, all linehaulers must be served before the first backhauler may be served. The VRPB is NP-hard, since it is a generalization of the VRP with $B = \emptyset$.

Another variant is the VRP with pickup and delivery (VRPPD), see for instance Desaulniers et al. [25]. Here a homogeneous or inhomogeneous fleet of vehicles based at one or several depots must satisfy a set of given transport requests. Each request is characterized by a pickup and a delivery vertex, and a (possibly negative) demand to be transported between them (for example, commodities or passengers). The VRPPD is NP-hard, since it is a generalization of the VRP arising when the pickup and delivery vertex are both the depot, and the demand is non-negative.

In the next chapter we present a mathematical model for the integrated optimization of school and bus starting times and bus schedules (IOSANA). The model we are going to present is based on the VRPTW, but there are some new constraints that do not fit in this framework. These are the coupling aspects of the time windows. That means, the departure and arrival of one vehicle at some node is not independent from what is going on around, but dependent on the arrival time of certain other vehicles at other nodes. Coupling constraints occur in two substantially different forms: We distinguish internal and external coupling.

As a modeling basis we take the VRPTW and use the notation from the previous section for the sets, parameters, and variables. If the start of the service at some node $v \in V$ is coupled to the start of the service at some other node $w \in V, v \neq w$, then we speak of an *internal coupling* of time windows. Let $I \subseteq V \times V$ be the set of all internally coupled nodes such that node w has to be served at least $a_{vw} \in \mathbb{Q}$ time units after node v.

As a linear inequality, internal coupling for $(v, w) \in I$ can be formulated as follows:

$$t_v + a_{vw} \leq t_w. \tag{3.19}$$

If the start of the service at some node $v \in V$ is coupled to the start of some event not contained in the model so far, then we speak of an *external coupling* of time windows. Suppose E is a set of external events, and $J_1, J_2 \subseteq E \times V$ are the sets of all pairs of nodes and events that are externally coupled. For each external event $e \in E$ we introduce a variable $T_e \in \mathbb{Q}_+$. The starting time for each external event $e \in E$ is bounded:

$$\underline{T}_e \leq T_e \leq \overline{T}_e. \tag{3.20}$$

For each $(e, v) \in J_1$, node v has to be served at least $\underline{b}_{ev} \in \mathbb{Q}$ time units after e, and for each $(e, v) \in J_2$, node v has to be served at most $\overline{b}_{ev} \in \mathbb{Q}$ time units after event e.

External coupling for $(e, v) \in J_1$ can be formulated as the following linear inequalities:

$$T_e + \underline{b}_{ev} \leq t_v, \tag{3.21}$$

and for $(e, v) \in J_2$ as

$$T_e + \overline{b}_{ev} \geq t_v. \tag{3.22}$$

A generic bicriteria mixed-integer model based on the VRPTW and including internal and external couplings of time windows is of the following form:

$$
\begin{aligned}
\min \quad & ((3.16), (3.17)) \\
\text{subject to} \quad & (3.9), (3.12), \dots, (3.15), (3.19), (3.21), (3.22) \\
& (3.10), (3.20) \\
& x \in \{0, 1\}^{|A|}, K \in \mathbb{Z}_+, t \in \mathbb{Q}_+^{|V|-1}, T \in \mathbb{Q}_+^{|E|}.
\end{aligned}
\tag{3.23}
$$

Model (3.23) is based on the VRPTW, but due to its additional constraints on its time windows it can be classified as a rich vehicle routing problem. In the VRPTW all time windows are independent from each other, but in our model they are coupled. Thus we suggest the term *vehicle routing problem with coupled time windows* (VRPCTW) for this new model.

We now describe a variant of the VRPCTW that will be important for the integrated optimization IOSANA.

Within IOSANA it is required that the starting time variables t and T are integral: $t \in \mathbb{Z}_+^{|V|-1}, T \in \mathbb{Z}_+^{|E|}$. Moreover it is necessary that the starting times are in discrete time slots of a certain length. To this end, for every internal coupling $(v, w) \in I$ the parameters $i_{vw}^1, i_{vw}^2 \in \mathbb{Q}_+$ are given, and (3.19) is altered to

$$i_{vw}^1 \cdot t_v + a_{vw} \leq i_{vw}^2 \cdot t_w. \tag{3.24}$$

For every external coupling $(e, v) \in J_1$ the parameters $j_{ev}^{1,1}, j_{ev}^{1,2} \in \mathbb{Q}_+$ and for every $(e, v) \in J_2$ the parameters $j_{ev}^{2,1}, j_{ev}^{2,2} \in \mathbb{Q}_+$ are introduced in inequalities (3.21) and (3.22), respectively:

$$j_{ev}^{1,1} \cdot T_e + \underline{b}_{ev} \leq j_{ev}^{1,2} \cdot t_v, \tag{3.25}$$
$$j_{ev}^{2,1} \cdot T_e + \bar{b}_{ev} \geq j_{ev}^{2,2} \cdot t_v. \tag{3.26}$$

In this case we speak of the *vehicle routing problem with coupled integral time windows* (VRPCITW).

3.4 Notes on the Complexity of VRPCTW

The theoretical complexity of the VRPCTW depends on the existence of proper time windows (instead of singleton time windows, i.e., intervals consisting of a single value). Up to a special case described in Theorem 4, the VRPCTW is NP-hard, that is, the decision problem for the VRPCTW (VRdPCTW, for short) is NP-complete. For this, we turn (3.23) into a single-criteria problem by scaling the two objective functions (3.16) and (3.17) by arbitrary non-negative weights as in (2.11). The VRdPCTW is then stated as follows: Given an instance of the VRPCTW and a number $L \in \mathbb{Q}$, is there a feasible solution of the instance that satisfies all VRPCTW constraints and has an objective function value less than L?

A special case which is computationally easy occurs if all trip time windows are singleton, i.e., $\underline{t}_v = \bar{t}_v$.

Theorem 4 *The VRPCTW and the VRPCITW with singleton starting times of the trips can be solved in polynomial time.*

Proof. In the case of the VRPCTW or the VRPCITW with singleton starting time windows of the trips, the computation of a solution decomposes into three independent problems.

1. We first check if the starting time windows are compatible with the internal coupling constraints (3.19). This can be done in $O(|I|)$ time. If an infeasibility occurs for some $(v, w) \in I$, then a feasible solution for the given instance does not exist.

2. We then check whether a feasible assignment of starting times to the external events is possible or not. We distinguish between the VRPCTW and the VRPCITW and start with the latter.

   With $\underline{t}_v = \bar{t}_v$ and the integrality of T_e, we obtain from (3.21) that the starting time of each external event $e \in E$ must satisfy for every $(e, v) \in J_1$

   $$T_e \leq \left\lfloor \frac{j_{ev}^{1,2} \cdot t_v - \underline{b}_{ev}}{j_{ev}^{1,1}} \right\rfloor , \tag{3.27}$$

   and from (3.22) we get

   $$T_e \geq \left\lceil \frac{j_{ev}^{2,2} \cdot t_v - \bar{b}_{ev}}{j_{ev}^{2,1}} \right\rceil \tag{3.28}$$

   for every $(e, v) \in J_2$. Define

   $$S_e^1 := \bigcap_{v:(e,v)\in J_1} \left] -\infty, \left\lfloor \frac{j_{ev}^{1,2} \cdot t_v - \underline{b}_{ev}}{j_{ev}^{1,1}} \right\rfloor \right] , \tag{3.29}$$

   $$S_e^2 := \bigcap_{v:(e,v)\in J_2} \left[\left\lceil \frac{j_{ev}^{2,2} \cdot t_v - \bar{b}_{ev}}{j_{ev}^{2,1}} \right\rceil , \infty \right[, \tag{3.30}$$

   then an assignment of starting times to the schools is possible if and only if for every external event $e \in E$ the set $S_e := S_e^1 \cap S_e^2 \cap [\underline{T}_e, \bar{T}_e] \cap \mathbb{Z}$ is non-empty, which can be checked in polynomial time $O(|J_1| + |J_2|)$. Any element from S_e can then be taken as a starting time for event e.

   In the case of the VRPCTW (without integral starting times) we can argue in a similar way. The only difference is that there is no rounding in (3.27), (3.28), (3.29), and (3.30) and that S_e is defined as $S_e := S_e^1 \cap S_e^2 \cap [\underline{T}_e, \bar{T}_e]$.

3. The computation of the schedules of the vehicles in the VRPCTW reduces to an assignment problem. For the solution of assignment problems, several polynomial (and pseudo-polynomial) time algorithms are known (see Ahuja, Magnanti and Orlin [1], for instance).

Altogether the theorem is proven.   □

Theorem 5 *The VRdPCTW and the VRdPCITW remain NP-complete if $I = \emptyset$ and $J = \emptyset$.*

Proof. In case of $I = J = \emptyset$, i.e., neither internal nor external couplings, VRdPCTW and VRdPCITW reduce to the classical VRdPTW with static time windows. A proof for the NP-completeness of VRdPTW was given by Savelsbergh [63].   □

Corollary 6 *The VRPCTW and the VRPCITW are NP-hard.*

Another interesting special case occurs if all variables x_{vw} are fixed to their bounds (that is, the schedules for all vehicles are given in advance), and it only remains to compute starting times for the trips and the external events.

Theorem 7

1. *The VRdPCTW with fixed schedules of the vehicles can be solved in polynomial time.*

2. *The VRdPCITW with fixed schedules of the vehicles is NP-complete. It can be solved in pseudo-polynomial time.*

Proof. Suppose all binary variables x_{vw} are fixed to their bounds such that the inequalities (3.12), ..., (3.15) are satisfied, then it remains either to compute t and T that fulfill $\underline{t} \le t \le \overline{t}, \underline{T} \le T \le \overline{T}$, (3.9), (3.19), (3.21), (3.22), or to show that no such values exist.

1. The inequality system formed by these inequalities is an LP2, i.e., a linear program with at most two non-zeros per inequality. Solving a linear program is in general a problem of polynomial time complexity. Using the special LP2 structure and the algorithm of Hochbaum and Naor [41], we can find feasible solutions (or disprove their existence) in strong polynomial $O(mn^2 \log m)$ time, where $n := |V| + |E|$ is the number of variables and $m := |V| + |I| + |J_1| + |J_2|$ is the number of inequalities of the corresponding LP2 system.

2. The inequality system formed by these inequalities together with the integrality constraints is a monotone IP2, which can be solved in pseudo-polynomial time $O(mU)$ using the algorithm of Bar-Yehuda and Rawitz (see Theorem 1), where $m := |V| + |I| + |J_1| + |J_2|$ and $U := \max\{U_1, U_2\}, U_1 := \max\{\overline{T}_e - \underline{T}_e : e \in E\}, U_2 := \max\{\overline{t}_v - \underline{t}_v : v \in V\}$.

Altogether the theorem is proven. □

Chapter 4

Mathematical Models for IOSANA

In this chapter we present a mathematical model for the integrated planning problem of bus and school starting times and bus scheduling (IOSANA). It is formulated as an integer programming problem based on the VRPCITW. We first discuss in Section 4.1 what kind of input data is necessary to set up the model. The variables and their bounds are presented in Section 4.2. The model itself comes in several variants starting from a what we call the bicriteria (or basic) model in Section 4.3 to more evolved versions later in Section 4.4. For the use as a real-world planning tool, these models are not sufficient, because several details are still left out. These details are presented in Section 4.5.

4.1 Sets and Parameters

We assume that it is desired to study the integrated planning problem in a given area or region, and at a certain period of the day. In the sequel, we call this area a *county*, but it might also correspond to less than a county (such as a single city) or more (for example if a large bus company operates in several adjacent counties). The period of the day we are mainly interested in are the morning rush hours (which gives a time horizon from approximately 5:00 to 9:00 a.m.), so in general we only consider the school starting times (and not, for instance, school ending times). We start with a detailed description of the necessary input sets and parameters of the model, see Table 4.1 for an overview. For the moment we take all necessary values for granted. Later, in Chapter 7 we address the problem how this data is generated out of the given raw data.

Trips

Let $\mathcal{V}$ be the set of all passenger trips in the given county. A *passenger trip* (or *trip* for short) $t \in \mathcal{V}$ is a sequence of bus stops. To each bus stop an arrival and a departure

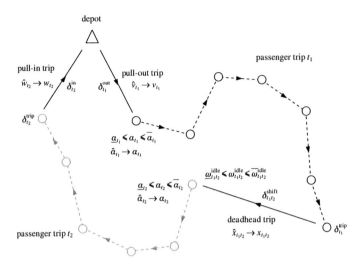

Figure 4.1: Two trips t_1, t_2 in a schedule.

time (of some bus) is assigned. The time difference between the departure at the first and the arrival at the last bus stop is called the *service duration*, and is denoted by $\delta_t^{\text{trip}} \in \mathbb{Z}_+$. (Generally all time-related parameters and variables in the model are integral with the unit "minute".) The current *starting time* of trip t, i.e., the departure time of a bus at the first bus stop in t, is given by $\hat{\alpha}_t \in \mathbb{Z}_+$. We assume that a time window $\underline{\alpha}_t, \overline{\alpha}_t \in \mathbb{Z}_+, \underline{\alpha}_t \leq \overline{\alpha}_t$ is given, in which the planned trip starting time must be. The trips in $\mathcal{V}$ play different roles in the transport of pupils to schools. The following types have to be distinguished: School trips, feeder and collector trips, and free trips, definitions of which are given in the subsequent.

The trips are served by vehicles. Usually these vehicles are buses, seldom trams or trains are in use, so in the sequel we simply speak of buses. The buses start and end their services at a *depot*. Denote $\mathcal{D}$ the set of all depots. In principle, every trip may be served by a new bus from some depot. The trip without passengers from the depot to the first bus stop of trip t is called *pull-out trip*. In the single-depot case, i.e., if $|\mathcal{D}| = 1$, denote $\delta_t^{\text{out}} \in \mathbb{Z}_+$ the time for a bus to drive this pull-out trip. Otherwise denote $\delta_{dt}^{\text{out}} \in \mathbb{Z}_+$ the driving time from depot $d \in \mathcal{D}$. When the bus arrives at the last bus stop of passenger trip t, it is either sent on the *pull-in trip*, i.e., back to the depot, or it is re-used to serve another passenger trip. The duration of the pull-in trip is denoted by $\delta_t^{\text{in}} \in \mathbb{Z}_+$ (single-depot case) or $\delta_{td}^{\text{in}} \in \mathbb{Z}_+$ (multi-depot case).

Instead of sending the bus back to the depot after having served a trip it is of course more sensible to re-use the bus to serve other trips, as long as this is possible. Thus we seek such a connection of trips. Let the set $\mathcal{A} \subset \mathcal{V} \times \mathcal{V}$ contain all pairs of trips (t_1, t_2)

that can in principle be connected. The intermediate trip from the last bus stop of trip t_1 to the first bus stop of trip t_2, where no passengers are transported, is called a *shift* or a *deadhead trip*. The duration of the deadhead trip is given by $\delta^{\text{shift}}_{t_1 t_2} \in \mathbb{Z}_+$. In order to absorb possible delays (due to traffic jams, for example), a minimum waiting time $\underline{\omega}^{\text{idle}}_{t_1 t_2} \in \mathbb{Z}_+$, typically 2 minutes, can be imposed after the deadhead trip, before starting the next passenger trip. In principle a maximum waiting time $\overline{\omega}^{\text{idle}}_{t_1 t_2} \in \mathbb{Z}_+, \underline{\omega}^{\text{idle}}_{t_1 t_2} \leq \overline{\omega}^{\text{idle}}_{t_1 t_2}$, can also be specified if the bus company wants to avoid long standing times of their buses. In practice this has not happened so far, so a typical value here is ∞ (infinity).

The connection of a pull-out trip, several passenger and intermediate deadhead trips and a final pull-in trip which are to be served by one and the same bus is called a *block* or *schedule* (see Figure 4.1). The current schedules of all vehicles are given by $\hat{v}_t, \hat{w}_t \in \{0,1\}$ for all $t \in \mathcal{V}$ and $\hat{x}_{t_1 t_2} \in \{0,1\}$ for all $(t_1, t_2) \in \mathcal{A}$. It is $\hat{v}_t = 1$ if and only if trip t is the first one in some schedule, $\hat{w}_t = 1$ if and only if trip t is the last one, and $\hat{x}_{t_1 t_2} = 1$ if and only if trip t_2 is currently served directly after trip t_1 by the same bus.

The parameters $\hat{\tau}, \hat{\alpha}, \hat{v}, \hat{w}, \hat{x}$ together are the *current solution* for the integrated planning problem. These values encode the today's starting times of school and buses and the today's schedules of the buses.

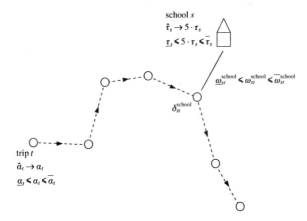

Figure 4.2: School trip t for school s.

Schools

Let $\mathcal{S}$ be the set of all schools in the county under consideration. The current starting time for school $s \in \mathcal{S}$ is given by $\hat{\tau}_s \in \mathbb{Z}_+$. It is allowed to change this starting time within some time window $\underline{\tau}_s, \overline{\tau}_s \in \mathbb{Z}_+, \underline{\tau}_s \leq \overline{\tau}_s$. Usually, this time window reflects the legal bounds on the school starting time (from 7:30 to 8:30 a.m.). Each school has some bus stops (usually, this is exactly one bus stop), where pupils get off the bus and walk the rest of the way to school.

The set $\mathcal{P} \subset \mathcal{S} \times \mathcal{V}$ consists of pairs (s, t), where trip t transports pupils to a bus stop of school s. In this case we say, t is a *school trip* for s (see Figure 4.2). The number of transported pupils by this trip is $\varphi_{st}^{\text{school}} \in \mathbb{Z}_+$. The time difference between the departure at the first bus stop of t and the arrival at the bus stop of s is denoted by $\delta_{st}^{\text{school}} \in \mathbb{Z}_+$. There is another time window for the pupils $\underline{\omega}_{st}^{\text{school}}, \overline{\omega}_{st}^{\text{school}} \in \mathbb{Z}_+, \underline{\omega}_{st}^{\text{school}} \leq \overline{\omega}_{st}^{\text{school}}$, specifying the minimal and maximal waiting time relative to the school starting time. The lower bound $\underline{\omega}_{st}^{\text{school}}$ is chosen according to the walking time from the bus stop where the pupils are dropped off, whereas the upper bound $\overline{\omega}_{st}^{\text{school}}$ is due to law restrictions. A typical time window is 5 – 45 minutes.

Let $\mathcal{C} \subset \mathcal{V} \times \mathcal{V}$ be the set of pairs (t_1, t_2), where t_1 is a trip that transports pupils to a so-called *changing bus stop*, where they leave the bus and transfer to trip t_2. We say, t_1 is a *feeder trip* for t_2 and, vice versa, t_2 is a *collector trip* for t_1 (see Figure 4.3). The number of transferring pupils between t_1 and t_2 is $\varphi_{t_1 t_2}^{\text{change}} \in \mathbb{Z}_+$. The driving time from the first bus stop of feeder trip t_1 to the changing bus stop is denoted by $\delta_{t_1 t_2}^{\text{feeder}} \in \mathbb{Z}_+$. For the collector trip, the corresponding parameter is $\delta_{t_1 t_2}^{\text{collector}} \in \mathbb{Z}_+$. At the changing bus stop, a time window $\underline{\omega}_{t_1 t_2}^{\text{change}}, \overline{\omega}_{t_1 t_2}^{\text{change}} \in \mathbb{Z}_+, \underline{\omega}_{t_1 t_2}^{\text{change}} \leq \overline{\omega}_{t_1 t_2}^{\text{change}}$ for the minimal and maximal waiting time is given. Typically, this time window is 0 – 10 minutes.

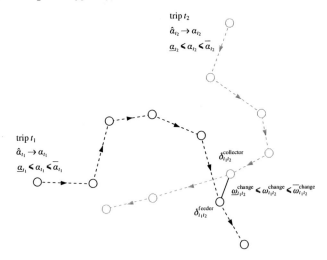

Figure 4.3: A feeder trip t_1 and a collector trip t_2 with $(t_1, t_2) \in \mathcal{C}$.

Note that a trip can have more than one type, p.e., it can be both a school and feeder

Table 4.1: Input data: sets and parameters

$\mathcal{D}$	$\ni d$	depots
$\mathcal{V}$	$\ni t$	bus trips (nodes)
$\mathcal{A}$	$\ni (t_1, t_2)$	connectable trips (arcs)
$\mathcal{S}$	$\ni s$	schools
$\mathcal{P}$	$\ni (s, t)$	school-trip pairings
$\mathcal{C}$	$\ni (t_1, t_2)$	feeder and collector trip pairings
$\hat{\tau}_s$		current school starting time
$\hat{\alpha}_t$		current trip starting time
$\hat{v}_t$		current first trip in block (pull-out trip)
$\hat{w}_t$		current last trip in block (pull-in trip)
$\hat{x}_{t_1 t_2}$		currently connected trips in block (deadhead trip)
δ_t^{trip}		time for serving entire trip
$\delta_{st}^{\text{school}}$		time for serving trip from start to school
$\delta_{t_1 t_2}^{\text{feeder}}$		time for feeder trip from start to changing b.s.
$\delta_{t_1 t_2}^{\text{collector}}$		time for collector trip from start to changing b.s.
δ_t^{out}		time for pull-out (single-depot)
δ_t^{in}		time for pull-in trip (single-depot)
δ_{td}^{out}		time for pull-out trip (multi-depot)
δ_{td}^{in}		time for pull-in trip (multi-depot)
$\delta_{t_1 t_2}^{\text{shift}}$		time for deadhead trip (single-depot)
$\delta_{t_1 t_2 d}^{\text{shift}}$		time for deadhead trip (multi-depot)
$\underline{\tau}_s, \overline{\tau}_s$		bounds on school starting time (lower, upper)
$\underline{\alpha}_t, \overline{\alpha}_t$		bounds on trip starting time
$\underline{\omega}_{st}^{\text{school}}, \overline{\omega}_{st}^{\text{school}}$		bounds on waiting time for pupils at school
$\underline{\omega}_{t_1 t_2}^{\text{change}}, \overline{\omega}_{t_1 t_2}^{\text{change}}$		bounds on waiting time at changing bus stop
$\underline{\omega}_{t_1 t_2}^{\text{idle}}, \overline{\omega}_{t_1 t_2}^{\text{idle}}$		minimum waiting time for bus after deadhead trip
$\varphi_{st}^{\text{school}}$		number of pupils on school bus
$\varphi_{t_1 t_2}^{\text{change}}$		number of transferring pupils between trips
$\underline{\kappa}_d, \overline{\kappa}_d$		bounds on the number of vehicles in depot (multi-depot)

trip. All trips that are neither school, feeder nor collector trips are called *free trips*. Free trips are obviously not important for the transport of pupils. However, they also have to be served by some bus. The time windows for these trips are usually quite narrow, for example plus/minus 15 minutes relative to the current starting time of the respective trip.

4.2 Variables and Bounds

The model contains binary and integer variables, see Table 4.2 for an overview.

For every trip $t \in \mathcal{V}$ the decision variables $v_t, w_t \in \{0, 1\}$ indicate if trip t is the first or

Table 4.2: Variables

v_t	$\in \{0,1\}$	first trip in block (single-depot)
w_t	$\in \{0,1\}$	last trip in block (single-depot)
$x_{t_1 t_2}$	$\in \{0,1\}$	connected trips in block (single-depot)
$\tilde{v}_t^d$	$\in \{0,1\}$	first trip in block (multi-depot)
$\tilde{w}_t^d$	$\in \{0,1\}$	last trip in block (multi-depot)
$\tilde{x}_{t_1 t_2}^d$	$\in \{0,1\}$	connected trips in block (multi-depot)
τ_s	$\in \mathbb{Z} \cap [\frac{1}{5}\underline{\tau}_s, \frac{1}{5}\overline{\tau}_s]$	school starting time (time slot)
α_t	$\in \mathbb{Z} \cap [\underline{\alpha}_t, \overline{\alpha}_t]$	trip starting time
$\omega_{st}^{\text{school}}$	$\in \mathbb{Z} \cap [\underline{\omega}_{st}^{\text{school}}, \overline{\omega}_{st}^{\text{school}}]$	waiting time for pupils at school
$\omega_{t_1 t_2}^{\text{change}}$	$\in \mathbb{Z} \cap [\underline{\omega}_{t_1 t_2}^{\text{change}}, \overline{\omega}_{t_1 t_2}^{\text{change}}]$	waiting time for pupils at changing bus stop
$\omega_{t_1 t_2}^{\text{idle}}$	$\in \mathbb{Z} \cap [\underline{\omega}_{t_1 t_2}^{\text{idle}}, \overline{\omega}_{t_1 t_2}^{\text{idle}}]$	waiting time for bus after deadhead trip
Δ_t^v	$\in \{0,1\}$	pull-out trip changed?
Δ_t^w	$\in \{0,1\}$	pull-in trip changed?
$\Delta_{t_1 t_2}^x$	$\in \{0,1\}$	deadhead trip changed?
Δ_s^τ	$\in \mathbb{Z}$	absolute value of the school starting time change
Δ_t^α	$\in \mathbb{Z}$	absolute value of trip starting time change

the last trip in some block, respectively. For every pair of trips $(t_1, t_2) \in \mathcal{A}$ the variable $x_{t_1 t_2} \in \{0,1\}$ indicates if t_1 and t_2 are in sequence in some block, that is, the same bus serves trip t_2 directly after finishing trip t_1 (apart from the deadhead trip and some idle time).

For every trip $t \in \mathcal{V}$ we introduce an integer variable $\alpha_t \in \mathbb{Z}$ representing its planned starting time, i.e., the departure of a bus at the first bus stop. Bounds on these variables are given by the corresponding time windows,

$$\underline{\alpha}_t \leq \alpha_t \leq \overline{\alpha}_t. \tag{4.1}$$

The school starting time is required to be in discrete time slots of 5 minutes (7:30, 7:35, 7:40, etc.). For every school $s \in \mathcal{S}$ we introduce an integer variable $\tau_s \in \mathbb{Z}$ with

$$\underline{\tau}_s \leq 5 \cdot \tau_s \leq \overline{\tau}_s. \tag{4.2}$$

Thus the planned school starting time of s is given by $5 \cdot \tau_s$.

For every $(s, t) \in \mathcal{P}$ the variable $\omega_{st}^{\text{school}} \in \mathbb{Z}$ keeps track of the waiting time for the pupils at the school bus stop relative to the school starting time. These variables are also bounded by time windows,

$$\underline{\omega}_{st}^{\text{school}} \leq \omega_{st}^{\text{school}} \leq \overline{\omega}_{st}^{\text{school}}. \tag{4.3}$$

The waiting time for pupils at the changing bus stop is settled by the variable $\omega_{t_1 t_2}^{\text{change}} \in \mathbb{Z}$ for every $(t_1, t_2) \in \mathcal{C}$. These variables are bounded from below and above,

$$\underline{\omega}_{t_1 t_2}^{\text{change}} \leq \omega_{t_1 t_2}^{\text{change}} \leq \overline{\omega}_{t_1 t_2}^{\text{change}}. \tag{4.4}$$

If two trips $(t_1, t_2) \in \mathcal{A}$ are connected by a deadhead trip, then there might be some waiting time for the bus driver before the start of trip t_2. For this, we introduce variables $\omega_{t_1 t_2}^{\text{idle}} \in \mathbb{Z}$ for all $(t_1, t_2) \in \mathcal{A}$ with

$$\underline{\omega}_{t_1 t_2}^{\text{idle}} \leq \omega_{t_1 t_2}^{\text{idle}} \leq \overline{\omega}_{t_1 t_2}^{\text{idle}}. \tag{4.5}$$

An assignment of values to the variables τ, α, v, w, x is called a *planned solution*. This is the way the buses are suggested to drive in the future (by solving one of the models below). Five different Δ variables are used (in the extended model) to measure the changes inbetween the current and the planned solution. The variables $\Delta_t^v, \Delta_t^w \in \{0, 1\}$ for all $t \in \mathcal{V}$ and $\Delta_{t_1 t_2}^x \in \{0, 1\}$ for all $(t_1, t_2) \in \mathcal{A}$ equal 1 if a current pull-out, pull-in, or deadhead trip (respectively) is no longer used in the planned solution, or vice versa. Otherwise the corresponding variables equal 0. (Note that we do not require these variables to be integral or binary. In a feasible solution this will turn out automatically due to the constraints.) Finally, the variables $\Delta_s^\tau \in \mathbb{Z}_+$ for all $s \in \mathcal{S}$ and $\Delta_t^\alpha \in \mathbb{Z}_+$ for all $t \in \mathcal{V}$ measure the absolute value of the time shift between current and planned school and bus starting time, respectively.

4.3 The Bicriteria Model

In its simplest version the model uses only parts of the input figures and variables described in the two previous sections. Parameters and variables are linked together in the constraints of the model, which are presented in Section 4.3.1. If there is more than one feasible solution to the model, a "best" one is wanted, where "best" is measured by a bicriteria objective function discussed in Section 4.3.2.

4.3.1 Constraints

Each trip is served by exactly one bus. That means, trip $t_2 \in \mathcal{V}$ either has a unique predecessor or it is the first one in some block:

$$\sum_{t_1 : (t_1, t_2) \in \mathcal{A}} x_{t_1 t_2} + v_{t_2} = 1. \tag{4.6}$$

Moreover, every trip $t_1 \in \mathcal{V}$ either has a unique successor or it is the last one in some block:

$$\sum_{t_2 : (t_1, t_2) \in \mathcal{A}} x_{t_1 t_2} + w_{t_1} = 1. \tag{4.7}$$

If trips $(t_1, t_2) \in \mathcal{A}$ are connected, then trip t_2 can only start after the bus has finished trip t_1, shifted from the end of t_1 to the start of t_2, and has waited a specified time to absorb possible delays. Additional waiting is permitted within certain limits if the bus arrives before the start of t_2. Using a sufficiently big value for M, these constraints can be formulated as linear inequalities:

$$\begin{aligned}
\alpha_{t_1} + \delta_{t_1}^{\text{trip}} + \delta_{t_1 t_2}^{\text{shift}} + \underline{\omega}_{t_1 t_2}^{\text{idle}} - M \cdot (1 - x_{t_1 t_2}) &\leq \alpha_{t_2}, \\
\alpha_{t_1} + \delta_{t_1}^{\text{trip}} + \delta_{t_1 t_2}^{\text{shift}} + \overline{\omega}_{t_1 t_2}^{\text{idle}} + M \cdot (1 - x_{t_1 t_2}) &\geq \alpha_{t_2}.
\end{aligned} \tag{4.8}$$

For the moment, it is enough to think of M as a large but finite number. In fact, M actually depends on $(t_1, t_2) \in \mathcal{A}$, so every inequality has its own $M_{t_1 t_2}$. Later in Section 6.1.1 we compute best possible values for $M_{t_1 t_2}$.

For every $(t_1, t_2) \in \mathcal{C}$ the starting times of both bus trips must be synchronized in such way that t_2 arrives at the changing bus stop after trip t_1 within a small time window. This is assured by the following inequalities:

$$\begin{aligned}
\alpha_{t_1} + \delta_{t_1 t_2}^{\text{feeder}} + \underline{\omega}_{t_1 t_2}^{\text{change}} &\leq \alpha_{t_2} + \delta_{t_1 t_2}^{\text{collector}}, \\
\alpha_{t_1} + \delta_{t_1 t_2}^{\text{feeder}} + \overline{\omega}_{t_1 t_2}^{\text{change}} &\geq \alpha_{t_2} + \delta_{t_1 t_2}^{\text{collector}}.
\end{aligned} \tag{4.9}$$

For every $(s, t) \in \mathcal{P}$ the starting times of trip t and school s have to be chosen such that the waiting time restrictions for the pupils at school s are met. Thus, we add the following inequalities to the model in order to synchronize the start of bus trips and schools:

$$\begin{aligned}
\alpha_t + \delta_{st}^{\text{school}} + \underline{\omega}_{st}^{\text{school}} &\leq 5 \cdot \tau_s, \\
\alpha_t + \delta_{st}^{\text{school}} + \overline{\omega}_{st}^{\text{school}} &\geq 5 \cdot \tau_s.
\end{aligned} \tag{4.10}$$

Regarding our introduction of the VRPCTW problem class in the previous chapter, constraints (4.9) can be seen as internal couplings and constraints (4.10) as external couplings, where the school starting time is the external event.

4.3.2   Objective Functions

The optimization basically has two goals: Finding a minimum number of buses to serve all trips, and deploying these buses in the most efficient way, that is, minimize the sum of all deadhead trips. Thus as objective functions we have:

- The total number of deployed vehicles:

$$\sum_{t \in \mathcal{V}} v_t, \tag{4.11}$$

- and the driving times of all pull-out, pull-in, and deadhead trips:

$$\sum_{t \in \mathcal{V}} \delta_t^{\text{out}} \cdot v_t + \sum_{(t_1, t_2) \in \mathcal{A}} \delta_{t_1 t_2}^{\text{shift}} \cdot x_{t_1 t_2} + \sum_{t \in \mathcal{V}} \delta_t^{\text{in}} \cdot w_t. \tag{4.12}$$

Summing up we have the following bicriteria integer optimization problem:

$$
\begin{array}{rl}
\min & ((4.11), (4.12)) \\
\text{subject to} & (4.6), \ldots, (4.10) \\
& (4.1), (4.2) \\
& v, w \in \{0,1\}^{|\mathcal{V}|}, x \in \{0,1\}^{|\mathcal{A}|} \\
& \tau \in \mathbb{Z}^{|\mathcal{S}|}, \alpha \in \mathbb{Z}^{|\mathcal{V}|}.
\end{array}
\tag{4.13}
$$

4.4 The Multicriteria Model

In its basic version above the model has a bicriteria objective function with an emphasis only on the reduction of the number of deployed vehicles and their efficient deployment in terms of deadhead trips. However, in the planning process it is desired to have a better control on some quality-of-service aspects of the final solution. Thus the model has to be enriched with a more detailed multicriteria objective function so that these goals are additionally taken into consideration. We take the bicriteria model (4.13) as a basis and describe in Section 4.4.1 and Section 4.4.2 the necessary changes of the constraints and objective functions, respectively.

4.4.1 Constraints

The new constraints can be grouped into two sections: First, constraints that track the various waiting times, and second, constraints that measure the differences between current and planned solution.

Tracking the Waiting Times

In the bicriteria model inequalities (4.8) limit the minimum and maximum waiting time of a bus between two trips $(t_1, t_2) \in \mathcal{A}$ which are connected in some schedule. The actual waiting time is now incorporated in these inequalities as an additional variable:

$$
\begin{array}{rl}
\alpha_{t_1} + \delta_{t_1}^{\text{trip}} + \delta_{t_1 t_2}^{\text{shift}} + \omega_{t_1 t_2}^{\text{idle}} - M \cdot (1 - x_{t_1 t_2}) & \leq \alpha_{t_2}, \\
\alpha_{t_1} + \delta_{t_1}^{\text{trip}} + \delta_{t_1 t_2}^{\text{shift}} + \omega_{t_1 t_2}^{\text{idle}} + M \cdot (1 - x_{t_1 t_2}) & \geq \alpha_{t_2}.
\end{array}
\tag{4.14}
$$

For trips $(t_1, t_2) \in \mathcal{C}$ inequalities (4.9) are synchronizing the waiting time for pupils at a changing bus stop. The precise waiting time is recorded with the following equalities:

$$
\alpha_{t_1} + \delta_{t_1 t_2}^{\text{feeder}} + \omega_{t_1 t_2}^{\text{change}} = \alpha_{t_2} + \delta_{t_1 t_2}^{\text{collector}}.
\tag{4.15}
$$

For every school s and every trip t with $(s,t) \in \mathcal{P}$ the waiting times for pupils at their school was limited by inequalities (4.10). These waiting times are now recorded in an additional variable by these equalities:

$$\alpha_t + \delta_{st}^{\text{school}} + \omega_{st}^{\text{school}} = 5 \cdot \tau_s. \tag{4.16}$$

Differences between Current and Planned Solution

For recording the differences between current and planned schedules we add the following inequalities to the model. For all pull-out trips $t \in \mathcal{V}$ let

$$\begin{aligned}
\hat{v}_t - v_t &\leq \Delta_t^v, \\
v_t - \hat{v}_t &\leq \Delta_t^v,
\end{aligned} \tag{4.17}$$

for all pull-in trip $t \in \mathcal{V}$ let

$$\begin{aligned}
\hat{w}_t - w_t &\leq \Delta_t^w, \\
w_t - \hat{w}_t &\leq \Delta_t^w,
\end{aligned} \tag{4.18}$$

and for all deadhead trips $(t_1, t_2) \in \mathcal{A}$ let

$$\begin{aligned}
\hat{x}_{t_1 t_2} - x_{t_1 t_2} &\leq \Delta_{t_1 t_2}^x, \\
x_{t_1 t_2} - \hat{x}_{t_1 t_2} &\leq \Delta_{t_1 t_2}^x.
\end{aligned} \tag{4.19}$$

The relative time shifts of school and trip starting times are transformed into absolute values by the following inequalities. For all schools $s \in \mathcal{S}$ we let

$$\begin{aligned}
\hat{\tau}_s - 5 \cdot \tau_s &\leq \Delta_s^\tau, \\
5 \cdot \tau_s - \hat{\tau}_s &\leq \Delta_s^\tau,
\end{aligned} \tag{4.20}$$

and for all passenger trips $t \in \mathcal{V}$ we let

$$\begin{aligned}
\hat{\alpha}_t - \alpha_t &\leq \Delta_t^\alpha, \\
\alpha_t - \hat{\alpha}_t &\leq \Delta_t^\alpha.
\end{aligned} \tag{4.21}$$

We remark that all above inequalities (4.17), ..., (4.21) only work properly if the corresponding Δ variables are to be minimized by the objective function (see next section), which is a reasonable assumption for practical purposes .

4.4.2 Objective Functions

The additional variables can be seen as shape variables which give us a much more refined control on the actual solution of the model. Instead of two there are now eight goals for the optimization. These additional goals are:

- the waiting times of buses between two trips, after the deadhead trip:

$$\sum_{(t_1,t_2)\in\mathcal{A}} \omega_{t_1 t_2}^{\text{idle}}, \tag{4.22}$$

- the waiting times for pupils at their schools:

$$\sum_{(s,t)\in\mathcal{P}} \varphi_{st}^{\text{school}} \cdot \omega_{st}^{\text{school}}, \tag{4.23}$$

- the waiting times at the transfer bus stops:

$$\sum_{(t_1,t_2)\in\mathcal{C}} \varphi_{t_1 t_2}^{\text{change}} \cdot \omega_{t_1 t_2}^{\text{change}}, \tag{4.24}$$

- the absolute change of the school starting times:

$$\sum_{s\in\mathcal{S}} \Delta_s^{\tau}, \tag{4.25}$$

- the absolute change of the starting times of the bus trips:

$$\sum_{t\in\mathcal{V}} \Delta_t^{\alpha}, \tag{4.26}$$

- and the difference between current and planned bus schedules:

$$\sum_{t\in\mathcal{V}} \Delta_t^{v} + \sum_{(t_1,t_2)\in\mathcal{A}} \Delta_{t_1 t_2}^{x} + \sum_{t\in\mathcal{V}} \Delta_t^{w}. \tag{4.27}$$

Now the multicriteria optimization problem can be stated as follows:

$$
\begin{aligned}
\min \quad & ((4.11),(4.12),(4.22),\ldots,(4.27)) \\
\text{subject to} \quad & (4.6),(4.7),(4.14),\ldots,(4.21) \\
& (4.2),\ldots,(4.5) \\
& v,w\in\{0,1\}^{|\mathcal{V}|}, x\in\{0,1\}^{|\mathcal{A}|} \\
& \tau\in\mathbb{Z}^{|\mathcal{S}|}, \alpha\in\mathbb{Z}^{|\mathcal{V}|} \\
& \omega^{\text{idle}}\in\mathbb{Z}^{|\mathcal{A}|}, \omega^{\text{school}}\in\mathbb{Z}^{|\mathcal{P}|}, \omega^{\text{change}}\in\mathbb{Z}^{|\mathcal{C}|} \\
& \Delta^v,\Delta^w\in\{0,1\}^{|\mathcal{V}|}, \Delta^x\in\{0,1\}^{|\mathcal{A}|} \\
& \Delta^{\tau}\in\mathbb{Z}^{|\mathcal{S}|}, \Delta^{\alpha}\in\mathbb{Z}^{|\mathcal{V}|}.
\end{aligned}
\tag{4.28}
$$

4.5 Additional Requirements

From the applicational point-of-view both the bicriteria and the multicriteria model described above provide only a basic framework, a reasonable starting point. For their application to real-world problem instances extensions are often necessary. These extensions are additional requirements of certain schools or restrictions to the starting times of some trips. From our experiences with some real-world instances we identified the following requirements and show how they can be incorporated in any of the models.

Pre-existing Deadhead Trips

In real life, deadhead trips are not always trips without passengers, as we defined them before. It sometimes occurs that passengers stay on the bus during a deadhead trip. In these cases the customers do not take notice of this deadhead trip. To them it seems to be a single, continuous trip instead of two separate ones. Since in our model all trips can be arbitrarily connected with each other, it is likely to happen that currently connected trips will afterwards be in different schedules. For the customers, this would mean that they have to change the bus. If this decrease in the quality-of-service is undesired then we can explicitly forbid this in the model by fixing the corresponding decision variables. If $\mathcal{A}' \subseteq \mathcal{A}$ is a subset of deadhead trips that have to be respected in the planned solution, then we add to the model the equalities $x_{t_1 t_2} = 1$ for all $(t_1, t_2) \in \mathcal{A}'$.

Inhomogeneous Fleet of Vehicles

So far all buses are identical in our models. In reality there are several types of buses, such as small buses with up to 100 seats, large buses with up to 150 seats, or air-conditioned long-distance buses. It is therefore not possible to assign every bus to every trip. Usually the operator (the bus company) decides which trip needs what kind of bus type. The easiest way to incorporate this into the model is the following: If trip t_1 needs some bus type and trip t_2 needs some other bus type, we let $x_{t_1 t_2} = x_{t_2 t_1} = 0$ to avoid a connection of t_1 with t_2 and vice versa. Another more sophisticated way to handle this constraint is shown at the end of this section (on page 49).

Competing Bus Companies

In some counties more than one bus company is operating. The trips are assigned to the companies, and in general it is not possible to make schedules out of trips which belong to different companies. If trip t_1 belongs to some company and trip t_2 belongs to some other company, we let $x_{t_1 t_2} = x_{t_2 t_1} = 0$, and thus avoid a connection of t_1 with t_2 and vice versa.

Cooperating Schools

Schools cooperate for various reasons. They share teachers, have partnerships for common lessons, or use a sports field or swimming pools at the same time. Of course, the optimization must not disturb those long-term cooperations. This is achieved by keeping the relative school starting times of all cooperating schools. Let $\mathcal{S}' \subseteq \mathcal{S}$ be a subset of cooperating schools, then we add the following constraint to the model:

$$5 \cdot \tau_{s_i} - 5 \cdot \tau_{s_{i+1}} = \hat{\tau}_{s_i} - \hat{\tau}_{s_{i+1}}, \quad \forall \, s_i, s_j \in \mathcal{S}', s_i \neq s_j. \tag{4.29}$$

Because of the transitivity of the equality relation we only have to add the equalities (4.29) for all $i = 1, \ldots, N - 1$ for an arbitrary ordering of the cooperating schools, $\mathcal{S}' = \{s_1, \ldots, s_N\}$.

Denote by $\mathcal{S}'_1, \ldots, \mathcal{S}'_C \subset \mathcal{S}$ (with $\mathcal{S}'_i \cap \mathcal{S}'_j = \emptyset$ for all $i, j \in \{1, \ldots, C\}, i \neq j$) the family of schools in different cooperations. Each subset $\mathcal{S}'_i, i \in \{1, \ldots, C\}$ is endowed with an arbitrary ordering of the elements so that we can write $s_1 \leq s_2$ with respect to this ordering.

Tacted Lines

Until now we considered the trips as independent from each other. Their starting times must only meet the waiting time requirements at school and change bus stops, but apart from that they are freely alterable. In reality we have to distinguish between *demand-oriented* and *tact-oriented* lines. So far the trips in our model belong to demand-oriented lines. This means, they are served at irregular intervals to meet a certain demand, such as pupils on their way to school or workers on their way to a factory. In the rush hour of the county's bigger cities, bus companies often decide to offer tact-oriented lines. That is, within a certain fixed period of time a bus is serving a trip of the line. Similar to cooperating schools the relative time difference within all trips of a tacted line must be preserved. Let $\mathcal{V}' \subseteq \mathcal{V}$ be a subset of trips belonging to a tacted line, and let $\mathcal{V}' = \{t_1, \ldots, t_N\}$ be an arbitrary ordering of the trips, then we add the following equalities to the model:

$$\alpha_{t_i} - \alpha_{t_{i+1}} = \hat{\alpha}_{t_i} - \hat{\alpha}_{t_{i+1}}, \quad \forall i = 1, \ldots, N - 1. \tag{4.30}$$

Denote by $\mathcal{V}'_1, \ldots, \mathcal{V}'_T \subset \mathcal{V}$ (with $\mathcal{V}'_i \cap \mathcal{V}'_j = \emptyset$ for all $i, j \in \{1, \ldots, T\}, i \neq j$) the family of trips in different tacted lines. Each subset $\mathcal{V}'_i, i \in \{1, \ldots, T\}$, is endowed with an arbitrary ordering of the elements so that we can write $t_1 \leq t_2$ with respect to this ordering.

Multiple Depots and Inhomogeneous Fleets (Revised)

So far we only considered the single-depot case where all trips start and end at the same depot. An extension to the multi-depot case is not straight-forward but a substantial modification of the model, because new families of variables have to be used.

Let $\mathcal{D}$ be a set of depots where vehicles start and end their duties. It is required that every schedule ends at the same depot where it started. We introduce additional variables $\tilde{v}^d, \tilde{w}^d \in \{0, 1\}^{|\mathcal{V}|}$ and $\tilde{x}^d \in \{0, 1\}^{|\mathcal{A}|}$ with an index representing the corresponding depot $d \in \mathcal{D}$. For $t \in \mathcal{V}$ and $d \in \mathcal{D}$ the binary variables $\tilde{v}_t^d$ and $\tilde{w}_t^d$ equal 1 if and only if trip t is served by a vehicle from depot d and t is the first or the last trip in some block, respectively. Similarly $\tilde{x}_{t_1 t_2}^d = 1$ if and only if trip t_1 is preceding trip t_2 and both are

served by a vehicle from depot d. The new variables are coupled to the previous ones
by the following constraints:

$$v_t = \sum_{d \in \mathcal{D}} \tilde{v}_t^d, \quad \forall\, t \in \mathcal{V}, \tag{4.31}$$

$$w_t = \sum_{d \in \mathcal{D}} \tilde{w}_t^d, \quad \forall\, t \in \mathcal{V}, \tag{4.32}$$

$$x_{t_1 t_2} = \sum_{d \in \mathcal{D}} \tilde{x}_{t_1 t_2}^d, \quad \forall\, (t_1, t_2) \in \mathcal{A}. \tag{4.33}$$

Extending the bicriteria model (4.13) to a multi-depot bicriteria model, the constraints
(4.6) and (4.7) saying that each trip is served by exactly one vehicle are no longer
sufficient. In addition we have to make sure that each vehicle returns to the depot
where it started from. This is done for each $(t_2, d) \in \mathcal{V} \times \mathcal{D}$ by the following equality
constraint:

$$\sum_{t_1 : (t_1, t_2) \in \mathcal{A}} \tilde{x}_{t_1 t_2}^d + \tilde{v}_{t_2}^d = \sum_{t_3 : (t_2, t_3) \in \mathcal{A}} \tilde{x}_{t_2 t_3}^d + \tilde{w}_{t_2}^d. \tag{4.34}$$

In addition, lower and upper bounds on the capacity of every depot $d \in \mathcal{D}$ can be
enforced:

$$\underline{\kappa}_d \le \sum_{t \in \mathcal{V}} \tilde{v}_t^d \le \overline{\kappa}_d, \tag{4.35}$$

where $\underline{\kappa}_d$ is the smallest number of vehicles that have to be in the depot d, and $\overline{\kappa}_d$
is the maximum number of vehicles. Depending on whether inequalities (4.35) is part
of the model or not, we speak of the capacitated or uncapacitated multi-depot model,
respectively.

As mentioned, the incorporation of multiple depots offers also another opportunity to
deal with an inhomogeneous fleet of vehicles. A disadvantage of the above method, where
simply all deadhead trips between trips with different types of vehicles were excluded,
occurs when the real-world situation is asymmetric. That is, when a small vehicle
cannot serve a trip where a large vehicle is needed, but a large vehicle can serve a trip
for a small one. Here the concept of multiple depots can be extended to the case of
an inhomogeneous fleet of vehicles, where d represents both the physical location of the
garage *and* the type of the available vehicle. Then different deadhead driving times per
vehicle can also be taken into consideration, since a large bus might be slower and has
to drive an other (longer) way compared to a smaller one.

The only objective function to be changed is (4.12). Here the pull-out and pull-in trips
from and to different depots and, in case of an inhomogeneous fleet, the different driving
times for deadhead trips have to be taken into account as follows:

$$\sum_{d \in \mathcal{D}} \left(\sum_{t \in \mathcal{V}} \delta_{td}^{\text{out}} \cdot \tilde{v}_t^d + \sum_{(t_1, t_2) \in \mathcal{A}} \delta_{t_1 t_2 d}^{\text{shift}} \cdot \tilde{x}_{t_1 t_2}^d + \sum_{t \in \mathcal{V}} \delta_{td}^{\text{in}} \cdot \tilde{w}_t^d \right). \tag{4.36}$$

Then, as an example, the capacitated multi-depot bicriteria model looks as follows:

$$
\begin{aligned}
\min \quad & ((4.11), (4.36)) \\
\text{subject to} \quad & (4.6), \ldots, (4.10), (4.31), \ldots, (4.35) \\
& (4.1), (4.2) \\
& \tilde{v}, \tilde{w} \in \{0,1\}^{|\mathcal{V}| \cdot |\mathcal{D}|}, \tilde{x} \in \{0,1\}^{|\mathcal{A}| \cdot |\mathcal{D}|} \\
& v, w \in \{0,1\}^{|\mathcal{V}|}, x \in \{0,1\}^{|\mathcal{A}|} \\
& \tau \in \mathbb{Z}^{|\mathcal{S}|}, \alpha \in \mathbb{Z}^{|\mathcal{V}|}.
\end{aligned}
\tag{4.37}
$$

However, neither the multi-depot case nor the case of different bus types was ever requested in the counties we studied so far, so we can refrain from their further discussion.

Chapter 5

Primal Aspects

In general, hard combinatorial optimization problems are solved by attacking them from two sides, called primal and dual side of the problem. The primal side discussed in this chapter is the one of feasible solutions. We are faced with the problem how to quickly generate good feasible solutions from scratch with a low objective function value (in the case of minimization problems). In literature, many heuristics and metaheuristics are described and applied to a wide range of different optimization problems. They are known by buzzwords such as greedy, local search, genetic algorithms, tabu search, simulated annealing, or ant colony optimization. In this chapter, we present a new metaheuristic for which we suggest the term *parametrized greedy heuristic* or *pgreedy*, for short. This new heuristic glues classical greedy construction elements with techniques from global optimization. In Section 5.1 we will present the central ideas of this new approach first by using the traveling salesman problem (TSP) and then in a more general setting for arbitrary mixed-integer problems. In Section 5.2 we discuss how pgreedy has to be tailored to become a heuristic for IOSANA.

5.1 How to Parameterize a Greedy Heuristic

Greedy-type heuristics are used in many special-purpose optimization software packages where a good feasible solution to a given instance of some (combinatorial) problem is required after a very short amount of time (typically, a few seconds). They are designed to construct feasible solutions to a given problem from scratch by step-by-step inserting always the best immediate, or local, solution while finding an answer to a given problem instance. To obtain in fact good solutions, the crucial point within every greedy algorithm is having a proper criterion that selects these local solutions and thus is responsible for the search direction. For some optimization problems greedy algorithms are able to find the globally optimal solution. For example, Dijkstra's algorithm actually is a greedy algorithm which finds a shortest path between two nodes in a given graph. On the other hand, there is no known greedy algorithm that finds a minimum Hamiltonian path, i.e., a solution to the TSP (or the ATSP).

5.1.1 An Introductory Example

In case of the ATSP a simple greedy heuristic is the *nearest-neighbour heuristic* (NN) that works as follows. Given is a complete graph $G = (V, A)$ with non-negative arc weights $c_{vw} \in \mathbb{Q}_+$. The salesman starts at an arbitrary node v_0 and then visits the node nearest to the starting node. That is, a node v_1 with $v_1 = \mathrm{argmin}\{s_{v_0 w} : (v_0, w) \in A\}$ is selected, where the scoring function s is defined as

$$s_{vw} := c_{vw}, \quad \forall \, (v, w) \in A. \tag{5.1}$$

From there the salesman visits the nearest node v_2 that was not visited so far. In general, in the k-th step of the heuristic we seek

$$v_k = \mathrm{argmin}\{s_{v_{k-1} w} : (v_{k-1}, w) \in A_k\} \tag{5.2}$$

and insert it into the tour. Here A_k is defined as $A_k := A_{k-1} \backslash \{(v_{k-1}, v_0), \dots, (v_{k-1}, v_{k-2})\}$ for all $k \geq 2$ and $A_1 := A$. These steps are iteratively repeated until all nodes are visited and the salesman returns to the start node v_0.

This might be the first heuristic that almost everyone comes up with. It is probably close to a practitioner's approach. However, the solutions found by this heuristic sometimes are of poor quality. The reason is, as one can see in Figure 5.1 below, that some nodes are "forgotten" during the course of the algorithm. They have to be inserted in later steps towards the end at relatively high costs. For the instance shown in Figure 5.1 the nearest neighbor heuristics results in a tour with a tour length of 677 units. (This instances was generated randomly, where 50 cities (or nodes) with integer coordinates were selected randomly on a 100×100 grid. For c_{vw} we take the euclidean distance v and w. In fact, the so-generated ATSP is a (symmetric) TSP instance, which for the heuristic presented in this section makes no difference.)

We extend now this greedy heuristic to a parametrized greedy heuristic (pgreedy) where the local criterion is a weighted combination of several sub-criteria. The crucial point for pgreedy is that more than a single criterion (as in greedy heuristics) is needed. In the k-th step of the nearest-neighbor heuristic (with $1 \leq k \leq |V|$), one can consider the "way back to the start node", i.e., $c_{v_k v_0}$ as an additional term in the scoring function. The question is how to weight between these two goals, the nearest neighbor on the one hand and the way back on the other. The easiest way to do this is to weight linearly and thus to extend the scoring function (5.1) as follows

$$s_{vw}(\lambda_1, \lambda_2) := \lambda_1 \cdot c_{vw} + \lambda_2 \cdot c_{wv_0}, \quad \forall \, (v, w) \in A, \tag{5.3}$$

where λ_1 and λ_2 are some scalar parameters. Since the scoring function (5.1) is a special case of (5.3) for $\lambda_1 := 1$ and $\lambda_2 := 0$, the previous solution is now one singular element

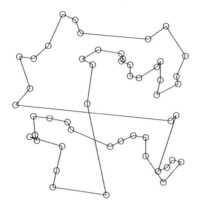

Figure 5.1: A TSP instance with 50 nodes, solved with NN.

within a larger solution family (parametrized by $\lambda \in \mathbb{Q}^2$). Within this family we expect to find better solutions. In fact, in case of the TSP instance shown in Figure 5.1 a better solution is obtained when using $\lambda_1 := 0.6$ and $\lambda_2 := -0.4$ as parameters in the scoring function (5.3). This solution (having a tour length of 592 units) is shown in Figure 5.2.

The question how to automatically find values for $\lambda \in \mathbb{Q}^2$ that lead to good solutions is addressed in detail in the following section. There we introduce techniques from global optimization where the parametrized greedy heuristic is called as a black-box function with parameters λ.

Before that, the generated solution still can be improved. This is done by a final local search step at the end of pgreedy. A well-known local search strategy is the so-called k-opt. A k-opt is an exchange of $k \in \mathbb{Z}, k \geq 2$ arcs of the solution. A recombination of this kind is accepted if the overall tour length is reduced, and rejected otherwise. k-opt steps are sequentially applied until no improvement is found anymore. For larger instances of the TSP one usually restricts to $k \in \{2, 3, 4\}$. For $k = 2$ we obtain after applying the k-opt procedure an improved solution with a tour length of 525 units, see Figure 5.3.

These three, a parametrized scoring function, a self-tuning algorithm to find the "right" parameters, and a local search procedure, are the main ingredients for a successful pgreedy heuristic. The details are presented in the following section.

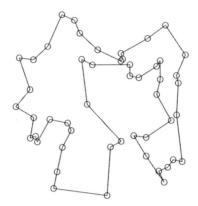

Figure 5.2: TSP instance, solved with pgreedy (without local search).

5.1.2   Greedy and PGreedy

We now leave the introductory description of parameterizing a greedy heuristic for the TSP to take a more general point of view to state and prove some theoretical results. Consider a bounded mixed-integer optimization problem

$$z^{\mathrm{opt}} = \min\{c^T x : Ax \le b, \underline{x} \le x \le \overline{x}, x \in \mathbb{Z}^{n_1} \times \mathbb{Q}^{n_2}\} \tag{5.4}$$

with $n_1 + n_2 = n, A \in \mathbb{Q}^{m \times n}, c \in \mathbb{Q}^n, b \in \mathbb{Q}^m, \underline{x}, \overline{x} \in \mathbb{Z}^{n_1} \times \mathbb{Q}^{n_2}$. Let $N^1 := \{1, \ldots, n_1\}, N^2 := \{n_1 + 1, \ldots, n\}$ be the index sets of the integer and the continuous variables, respectively, and let $N := N^1 \cup N^2$.

A *scoring function* is a mapping $s : N \times \mathbb{Q} \to \mathbb{Q} \cup \{\infty\}$ which yields a score $s(i, v)$ for the assignment of value $v \in \mathbb{Q}$ to variable x_i with index $i \in N$. Without loss of generality we assume that one is interested in finding $(i, v) \in N \times \mathbb{Q}$ that minimizes the scoring function. Then we can set $s(i, v) = \infty$ for infeasible assignments, including in particular the trivially infeasible ones, such as $i \in N^1$ and $v \in \mathbb{Q} \backslash \mathbb{Z}$, or $i \in N$ and $v \notin [\underline{x}_i, \overline{x}_i]$. For example, the scoring function (5.1) for the nearest neighbor heuristic for the TSP also is a scoring function in the just defined sense. To this end, let $s((i, j), 1) := c_{ij}$ for all $(i, j) \in A$ and $s((i, j), v) := \infty$ for all $v \in \mathbb{Q} \backslash \{1\}, (i, j) \in A$.

A *greedy heuristic* is defined as a procedure that selects in the k-th step an index $i \in N$ and a value $v \in \mathbb{Q}$ with $(i, v) = \operatorname{argmin}\{s(j, w) : j \in N_k, w \in \mathbb{Q}\}$ and sets $x_i^* := v$. If argmin is not unique, i.e., if there are two elements $(i_1, v_1), (i_2, v_2)$ with minimal score $s(i_1, v_1) = s(i_2, v_2)$, then we select the smaller of the two elements with respect to an

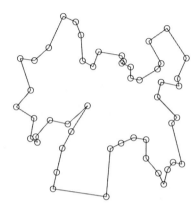

Figure 5.3: TSP instance, solved with pgreedy and local search.

arbitrary but fixed ordering on $N \times \mathbb{Q}$. (Note that N is finite and $\mathbb{Q}$ is countable, hence such ordering on $N \times \mathbb{Q}$ exists.) From the assignment of value v to variable x_i further restrictions on the range of other variables might be deduced due to the constraints. For example, if (A, b) contains a constraint of the form $\sum_{j \in J} x_j = 1$, where $J \subseteq N$ is some index set and $x \in \{0, 1\}^n$. Now assume that $(i, 1)$ is selected by the scoring function for some $i \in J$, then we set $x_i^* = 1$ and additionally can set $x_j^* = 0$ for all $j \in J \backslash \{i\}$. Denote $M_k \subseteq N_k$ the subset of additionally fixed variables. For the next round all variables with already assigned values are removed. For the index set of the remaining free variables we define $N_k := N_{k-1} \backslash (\{i\} \cup M_{k-1})$ for all $k \geq 2$ and $N_1 := N$. The algorithm terminates after at most n steps when there is no free (unassigned) variable left, i.e., $N_k = \emptyset$. An example for a greedy heuristic in this sense is the nearest neighbor heuristic for the TSP.

---

greedy$(A, c, b, \underline{x}, \overline{x}, s)$

Input:	optimization problem $A \in \mathbb{Q}^{n \times n}, c \in \mathbb{Q}^n, b \in \mathbb{Q}^m, \underline{x}, \overline{x} \in \mathbb{Z}^{n_1} \times \mathbb{Q}^{n_2}$
	scoring function $s : N \times \mathbb{Q} \to \mathbb{Q} \cup \{\infty\}$

(1)   Let $k := 1, N_1 := N$
(2)   **Repeat**
(3)   Let $(i_k, v_k) := \operatorname{argmin}\{s(i, v) : i \in N_k, v \in \mathbb{Q}\}$
(4)   Let $x_{i_k}^* := v_k$
(5)   Let $M_k \subseteq N_k$ be the set of additionally fixed variables
(6)   Select $x_i^* \in \mathbb{Z}$ for all $i \in M_k \cap N^1$ and $x_i^* \in \mathbb{Q}$ for all $i \in M_k \cap N^2$
(7)   Let $k := k + 1$
(8)   Let $N_k := N_{k-1} \backslash (\{i_k\} \cup M_{k-1})$
(9)   **Until** $N_k = \emptyset$

Output:	solution $x^* \in \mathbb{Z}^{n_1} \times \mathbb{Q}^{n_2}$

A greedy heuristic is called *adaptive* if it makes use of a family of scoring functions $s_k : N \times \mathbb{Q} \to \mathbb{Q}$, where the k-th scoring function depends on the fixed variables and the assigned values in the $k - 1$ steps before. For example, the nearest neighbor greedy heuristic for the TSP is not adaptive, because this procedure makes use of a single scoring function (5.1) that does not depend on selections done in previous steps. On the other hand, the greedy heuristic of Dijkstra is adaptive.

As one can imagine the actual selection of a proper scoring function s is essential for constructing a good greedy heuristic. The algorithm's output is a vector x^* called *solution*. A solution x^* is either feasible if $Ax^* \leq b, \underline{x} \leq x^* \leq \overline{x}, x^* \in \mathbb{Z}^{n_1} \times \mathbb{Q}^{n_2}$, or infeasible otherwise. A feasible solution is called *optimal* if for every other feasible solution y we have $c^T x^* \leq c^T y$. If a greedy heuristic always terminates with a feasible solution it is called *reliable*. If a reliable greedy heuristic always produces optimal solutions it is called *exact*.

A modification of the above greedy heuristics (adaptive or not) is the *enumerative greedy heuristic* or *egreedy* that for sure terminates with a feasible solution. Once an infeasibility of the solution under construction is detected (either during the construction phase or in the end) the index and value combinations that have led to this infeasibility are excluded and the search is continued from the point when the partial solution was still feasible (or at least where no infeasibility was detected). The information which indices and values are excluded during the search can be organized as a tree. The root node of the tree represents the start when all variables are unassigned and the construction starts. Each node of the tree represents an index and a value that is assigned to the corresponding variable. The tree is pruned at nodes where the assignment makes the partially constructed solution infeasible. For this we carry over the notion of feasibility for a solution to a partial solution under construction.

egreedy$(A, c, b, \underline{x}, \overline{x}, s)$	
Input:	optimization problem $A \in \mathbb{Q}^{m \times n}, c \in \mathbb{Q}^n, b \in \mathbb{Q}^m, \underline{x}, \overline{x} \in \mathbb{Z}^{n_1} \times \mathbb{Q}^{n_2}$
	scoring function $s : N \times \mathbb{Q} \to \mathbb{Q} \cup \{\infty\}$
(1)	Let $k := 1, N_1 := N, V_1 := N_1 \times \mathbb{Q}$
(2)	**Repeat**
(3)	Let $(i_k, v_k) = \operatorname{argmin}\{s(i, v) : (i, v) \in V_k\}$
(4)	Let $x_i^* := v_k$
(5)	Let $M_k \subseteq N_k$ be the set of additionally fixed variables
(6)	Select $x_j^* \in \mathbb{Z}$ for all $j \in M_k \cap N^1$ and $x_j^* \in \mathbb{Q}$ for all $j \in M_k \cap N^2$
(7)	**If** x^* is feasible **Then**
(8)	Let $k := k + 1$
(9)	Let $N_k := N_{k-1} \backslash (\{i_k\} \cup M_{k-1})$
(10)	Let $V_k := N_k \times \mathbb{Q}$
(11)	**Else**
(12)	Let $V_k := V_k \backslash \{(i_k, v_k)\}$
(13)	**If** $V_k = \emptyset$ **Then**
(14)	Let $k := k - 1$
(15)	**If** $k = 0$ **Then** Return "no solution exists"
(16)	Let $V_k := V_k \backslash \{(i_k, v_k)\}$
(17)	**End If**
(18)	**End If**
(19)	**Until** $N_k = \emptyset$
Output:	feasible solution $x^* \in \mathbb{Z}^{n_1} \times \mathbb{Q}^{n_2}$ or message "no solution exists"

The enumerative greedy heuristic has some aspects in common with the branch-and-bound method for solving mixed-integer problems: In fact, branchAndBound is an enumerative greedy heuristic with the additional capability to prune the search tree also at feasible nodes. Moreover, when branchAndBound terminates a globally optimal solution is returned (if the problem is feasible), whereas egreedy returns with the first feasible solution that was found. In both algorithms, branchAndBound and egreedy, it is desired to detect infeasibilities as early as possible to keep the search tree as small as possible. The difference between the two is that branchAndBound emphasizes a detection of a global optimal solution by enumerating large parts of the search space, whereas the *egreedy* emphasizes a quick construction of a feasible solution.

A *parametrized scoring function* with p parameters $\lambda \in \mathbb{Q}^p$ is a mapping $s : N \times \mathbb{Q} \times \mathbb{Q}^p \to \mathbb{Q}$ with argument (i, v, λ). In the sequel we restrict our discussion to the case of scoring functions that are linear in the parameters, i.e., for all $i \in N, v \in \mathbb{Q}, \lambda, \mu \in \mathbb{Q}^p$, and $t, u \in \mathbb{Q}$ we have $s(i, v, t \cdot \lambda + u \cdot \mu) = t \cdot s(i, v, \lambda) + u \cdot s(i, v, \mu)$. A greedy heuristic that makes use of such scoring function is hence called a *parametrized greedy heuristic* or *pgreedy*, for short. Suppose we have two pgreedy heuristics H^1, H^2 with parametrized scoring functions s^1, s^2 with p^1 and p^2 parameters, respectively, and $p^1 \leq p^2$. Then we say H^2 *includes* H^1 if $s^1 = s^2|_{N \times \mathbb{Q} \times \mathbb{Q}^{p_1}}$. For example, the pgreedy heuristic using (5.3) includes the nearest neighbor heuristic using (5.1), because $s_{vw} = s_{vw}(\lambda)$ for $\lambda = (1, 0)$.

The local selection of variables and values and hence the entire solution found by the pgreedy heuristic depends on the actual choice of $\lambda \in \mathbb{Q}^p$. We write $x^*(\lambda)$ for the solution found when the pgreedy algorithm is called with parameter λ, and $z^*(\lambda)$ for the corresponding objective function value, i.e., $z^*(\lambda) = c^T x^*(\lambda)$. We are now faced with the problem to find a vector λ^* with $z^*(\lambda^*) \leq z^*(\lambda)$ for all $\lambda \in \mathbb{Q}^p$ and hence to search for

$$z^{\mathrm{pgreedy}} = \min\{z^*(\lambda) : \lambda \in \mathbb{Q}^p\}. \tag{5.5}$$

If the domain for the parameters is bounded – and we show in Corollary 9 below that this can be assumed without loss of generality – then in principle a vector λ with weights that lead to the a feasible solution of (5.5) with a low objective function value can be found by sampling over a regular, sufficiently dense grid. For each λ of this grid the pgreedy heuristic has to be called and the best λ (i.e., the λ with the lowest objective function value $z^*(\lambda)$) is kept. However, in practice this approach turns out to be inefficient, even for a relatively small number of parameters ($p = 5$, for instance).

We use three different approaches, pure random search, hit-and-run, and improving hit-and-run. Later in Chapter 8 we evaluate their respective computational merits. The first and most simple idea is *random search*, where candidate λ parameters are selected randomly and independently from each other, until a maximum number of iterations is reached.

As a more sophisticated strategy, we use *improving hit-and-run* (*IHR* for short), an algorithm introduced by Zabinsky et al. [79] (see also Zabinsky [78]) to solve the global optimization problem (5.5). IHR is a randomized (Monte-Carlo) algorithm that automatically selects parameters which lead to good, possibly optimal solutions when used in a pgreedy heuristic. In a hybrid algorithm of pgreedy and IHR, a combination of a parametrized greedy algorithm and improving hit-and-run, IHR is used to compute the weights λ that take control of the parametrized scoring function and calls the pgreedy algorithm as a black-box to obtain a new objective function value. We now turn to the details.

The basic idea behind Improving Hit-and-Run is to use Hit-and-Run to generate candidate points randomly and accept only those that are an improvement with respect to the objective function. For a given objective function $z : B \to \mathbb{Q}$, defined on a bounded subset $B \subset \mathbb{Q}^p$ the IHR algorithm works as follows. We start with an initial $\lambda^0 \in B$, and set $k := 0$. The following steps are now repeated until a stopping criterion is met, for example, if the number of iterations k reaches a certain limit. Generate a random direction vector d^k uniformly distributed on the boundary of a unit hypersphere $\mathbb{S}^p := \{\lambda \in \mathbb{Q}^p : \|\lambda\|_2 = 1\}$. Generate a candidate point $w^{k+1} := \lambda^k + t \cdot d^k$, where t is generated by sampling uniformly over the line set $L_k := \{\lambda \in B : \lambda = \lambda^k + t \cdot d^k, t \in \mathbb{Q}_+\}$. If the candidate point is improving, i.e., $z(w^{k+1}) < z(\lambda^k)$, we set $\lambda^{k+1} := w^k$, otherwise $\lambda^{k+1} := \lambda^k$. (In the hit-and-run strategy, also non-improving points are accepted.) Finally, increase k by 1.

ihr(z, p)	
Input:	objective function $z : B \to \mathbb{Q}$
(1)	Let $k := 0, x_0 \in X$
(2)	**Repeat**
(3)	Generate a random direction vector $d^k \in \mathbb{S}^n$
(4)	Let $L^k := \{x \in X : x = x^k + \lambda d^k, \lambda > 0\}$
(5)	**If** $L^k \neq \emptyset$ **Then**
(6)	Generate a random candidate point $w \in L_k$
(7)	**If** $f(w) < f(x^k)$ **Then**
(8)	Let $k := k + 1$
(9)	Let $x^k := w$
(10)	**End If**
(11)	**End If**
(12)	**Until** a stop criterion is met
Output:	solution $\lambda \in \mathbb{Q}^p$

So far there is one major drawback for the application of IHR to obtain good (or optimal) parameters for pgreedy: The domain of parameters for the objective function has to be bounded to apply IHR.

The following observation is immediate but important on one hand to restrict the unbounded domain for the λ parameters to some bounded subset $B \subset \mathbb{Q}^p$. On the other hand, it is used to avoid the re-computation of the same solution over and over again.

Theorem 8 *Let $\lambda, \lambda' \in \mathbb{Q}^p$. If there is a positive scalar $t \in \mathbb{Q}_+$ such that $\lambda' = t \cdot \lambda$ then $x^*(\lambda) = x^*(\lambda')$ and hence $z^*(\lambda) = z^*(\lambda')$.*

Proof. Consider an arbitrary step k of the pgreedy heuristic with parameter λ. Let $(i, v) := \mathrm{argmin}\{s(j, w, \lambda) : j \in N, w \in \mathbb{Q}\}$ be the selected index and value in this step. That means, $s(i, v, \lambda) \leq s(j, w, \lambda)$ for all $j \in N, w \in \mathbb{Q}$. It follows that $t \cdot s(i, v, \lambda) \leq t \cdot s(j, w, \lambda)$. Since $s(i, v, \cdot)$ is linear, we get $s(i, v, t \cdot \lambda) \leq s(j, w, t \cdot \lambda)$ and hence $s(i, j, \lambda') \leq s(j, w, \lambda')$ for all $j \in N, w \in \mathbb{Q}$. Therefore $(i, v) = \mathrm{argmin}\{s(j, w, \lambda') : j \in N, w \in \mathbb{Q}\}$. Since k was arbitrarily chosen both heuristics select the same local best node in every step. Thus the constructed solutions and their objective function values are the same. $\square$

Corollary 9 *For every solution $x^*(\lambda)$ with $\lambda \in \mathbb{Q}^p \backslash \{0\}$ there is a $\lambda' \in [-1, 1]^p \backslash \{0\}$ such that $x^*(\lambda) = x^*(\lambda')$.*

Proof. Let $\lambda \in \mathbb{Q}^p \backslash \{0\}$. If already $\lambda \in [-1, 1]^p$ we are done. Otherwise set $t := 1/\max\{|\lambda_1|, \ldots, |\lambda_p|\}$ and let $\lambda' := t \cdot \lambda$, then $\lambda' \in [-1, 1]^p$ and $x^*(\lambda) = x^*(\lambda')$ by Theorem 8. $\square$

Because of Corollary 9 the IHR algorithm can be used to find optimal parameters for pgreedy. We have shown that

$$z^{\text{pgreedy}} = \min\{z^*(\lambda) : \lambda \in \mathbb{Q}^p\} = \min\{z^*(\lambda) : \lambda \in [-1,1]^p\}. \tag{5.6}$$

We can shrink the domain for the parameters even more.

Corollary 10 *Let* $\|\cdot\|$ *be an arbitrary norm. For every solution* $x^*(\lambda)$ *with* $\lambda \in \mathbb{Q}^p \backslash \{0\}$ *there is a* $\lambda' \in \mathbb{Q}^p$ *with* $\|\lambda'\| = 1$ *such that* $x^*(\lambda) = x^*(\lambda')$.

Proof. Let $\lambda \in \mathbb{Q}^p, \lambda \neq (0,0)$. If already $\|\lambda\| = 1$ then we have nothing to do anymore. Otherwise set $t := 1/\|\lambda\|$ and $\lambda' := t \cdot \lambda$. Then $\|\lambda'\| = 1$ and $x(\lambda) = x(\lambda')$ by Theorem 8. $\square$

In view of Corollary 10 we do not have to search the entire $\mathbb{Q}^p$ to evaluate (5.5), a proper subset of it is already sufficient. This will speed up computations considerably.

For example, consider a pgreedy with p parameters and the Euclidean norm $\|\lambda\| := (\lambda_1^2 + \ldots + \lambda_p^2)^{\frac{1}{2}}$. The set of parameters with $\|\lambda\| = 1$ is the p-dimensional unit hypersphere $\mathbb{S}^{p-1}$. A parametrization of this sphere is given by polar coordinates:

$$\Phi_p(\varphi) := \lambda := \begin{pmatrix} \cos \varphi_{p-1} \cdot \cos \varphi_{p-2} \cdot \ldots \cdot \cos \varphi_2 \cdot \cos \varphi_1 \\ \cos \varphi_{p-1} \cdot \cos \varphi_{p-2} \cdot \ldots \cdot \cos \varphi_2 \cdot \sin \varphi_1 \\ \cos \varphi_{p-1} \cdot \cos \varphi_{p-2} \cdot \ldots \cdot \cos \varphi_3 \cdot \sin \varphi_2 \\ \vdots \\ \cos \varphi_{p-1} \cdot \sin \varphi_{p-2} \\ \sin \varphi_{p-1} \end{pmatrix} \tag{5.7}$$

for $\varphi \in W^{p-1} := [-\pi, \pi[\times [-\frac{\pi}{2}, \frac{\pi}{2}[^{p-2}$ (see Kaballo [44] for the details).

5.2   PGreedy for IOSANA

Having introduced pgreedy with an example (TSP) and in an abstract way we are now ready to apply this new metaheuristic to solve instances of IOSANA. In this section we describe the necessary work that was done during the tailoring of pgreedy.

5.2.1   A Greedy Heuristic for IOSANA

From a theoretical point of view, we remarked in Section 3.4 that the VRPCTW and hence IOSANA is NP-hard. Checking feasibility with a fixed number of vehicles is NP-complete. Thus we cannot expect a polynomial algorithm for its solution unless $P = NP$.

In order to obtain good feasible solutions we now describe an application of the general greedy heuristics of the previous section to IOSANA. According to our description of greedy we need to construct an adequate scoring function that takes control of the search direction. For this it is necessary to take a closer look at the objective functions.

The Scoring Function

The IOSANA model exists in two different variants. First, we have the bicriteria model (4.13) of Section 4.3 and second, there is the multicriteria model (4.28) of Section 4.4. A first idea how to construct a scoring function is given by the objective functions of these two models. The most important goal is the reduction of the number of buses (4.11) and second, their efficient deployment (4.12). These two goals are lexicographically ordered, that is, (4.11) is (infinitely) more important than (4.12). All other goals (4.22), ..., (4.27) of the multicriteria model (4.28) take care of the quality-of-service. Since there are also hard restrictions in the model concerning the quality these additional objectives can be treated on a subordinate level. Thus the scoring function we are going to construct below is first taking care of the schedules and then, when the schedules of the vehicles are determined, the starting times are assigned to the schools and the trips. We describe here our scoring function in brief. The details follow in the next sections.

As in the case of the classical vehicle routing problem problem with fixed time windows (VRP) our scoring function for IOSANA favors an arc $(t_1^*, t_2^*) \in \mathcal{A}$ with minimal distance, i.e., $\delta_{t_1^* t_2^*}^{\text{shift}} \leq \delta_{t_1 t_2}^{\text{shift}}$ for all $(t_1, t_2) \in \mathcal{A}$. At the end, when there is no arc left and the schedules are determined, the scoring function selects proper starting times for the schools and the trips.

Here it is preferred that the new planned starting times are as close as possible to the current starting times so that there is a higher chance that the solution is accepted by the public (the pupils, the parents, teachers, etc., see our discussion in Section 1.4). Hence for the schools the scoring function selects a starting time $\tau_s^* \in \mathbb{Z} \cap [\underline{\tau}_s, \overline{\tau}_s]$ with minimal difference $|\tau^* - \hat{\tau}_s|$ to the current school starting time. The same idea applies for the trip starting times where a starting time $\alpha_t^* \in \mathbb{Z} \cap [\underline{\alpha}_t, \overline{\alpha}_t]$ with minimal difference $|\alpha^* - \hat{\alpha}_s|$ to the current trip starting time is selected by the scoring function. In view of the application (and the public opinion) the selection of school starting times is more crucial than the selection of the trip starting times. Hence the latter can be treated on the lowest level, after having settled the school starting times.

We now turn to the details of the schedule construction and starting time selection process. Note that due to our scoring function the entire solution process is split into two stages. In stage one a schedule of buses is computed such that preferably few buses are in use and the total length of all deadhead-trips is minimal. At this stage, it is not necessary to compute starting times for the schools and the trips. It only has to be assured that the schedules of the buses are feasible in such way that the starting times can be computed. Their actual computation according to the subordinate objectives is then done in stage two where the schedules from stage one are taken as fixed input values.

Deducing Further Restrictions by Constraint Propagation

In the definition of the greedy heuristic (on page 56) we already pointed out that after assigning a value to some selected variable, the deduction of further restriction on other variables is a crucial step. For the solution of VRPICTW (and IOSANA) instances, the corresponding procedure is the *starting time propagation*, which we describe now.

Before the first step of our heuristics starts working we have to check whether a feasible solution for a given instance of IOSANA exists or not. Mathematically this is done by solving an appropriate IP2 system. The steps described in the sequel are not only applied in the beginning but also later after every step in the construction phase of the heuristic. In principle, these steps are an application of the Bar-Yehuda and Rawitz algorithm (as presented in Section 2.2.5) to a subsystem of IOSANA, consisting of the variables α and τ, their trivial bounds (4.2), (4.1), and the inequalities (4.8), (4.9), (4.10), (4.29), (4.30).

There are two remarks in order. First, in the case of the multicriteria model the inequalities (4.8), (4.9), and (4.10) have to be replaced by the corresponding inequalities (4.14), (4.15), and (4.16). At first sight they do not fit the two-variable restriction of the IP2 framework (and its solution algorithms). However, in our case the additional ω variables are only slack variables. We can replace them by their respective lower and upper bounds to obtain an IP2. Once this IP2 is solved the ω variables can easily be reintroduced. The details are presented below in the derivation of (5.8).

Second, inequalities (4.8) and (4.14) have a binary variable x that also does not fit into the scheme. We remark that we do not take all of these inequalities into account but only those where x is already fixed to the upper bound 1. Only they can contribute to the starting time windows. Other inequalities where x is fixed to the lower bound 0 are "weak" and thus can be neglected.

School and trip starting times are coupled by the minimum and maximum waiting time restrictions of inequalities (4.10) or (4.16). The time window for the starting time of school s can be propagated onto the starting time window of trip t for all $(s,t) \in \mathcal{P}$. From (4.16) and the trivial constraints on the bounds of the variables (4.2) and (4.3) we obtain

$$\alpha_t = 5 \cdot \tau_s - \delta_{st}^{\text{school}} - \omega_{st}^{\text{school}} \leq 5 \cdot \tau_s - \delta_{st}^{\text{school}} - \underline{\omega}_{st}^{\text{school}} \leq \overline{\tau}_s - \delta_{st}^{\text{school}} - \underline{\omega}_{st}^{\text{school}}. \qquad (5.8)$$

We now compare the right-hand side of (5.8) with the previous upper bound $\overline{\alpha}_t$ on α_t. If it is less, a new *improved* upper bound is found. In general, we set for all $(s,t) \in \mathcal{P}$

$$\overline{\alpha}_t := \min\{\overline{\alpha}_t, \overline{\tau}_s - \delta_{st}^{\text{school}} - \underline{\omega}_{st}^{\text{school}}\}. \qquad (5.9)$$

In the same way, an improved lower bound can be derived for all $(s,t) \in \mathcal{P}$

$$\underline{\alpha}_t := \max\{\underline{\alpha}_t, \underline{\tau}_s - \delta_{st}^{\text{school}} - \overline{\omega}_{st}^{\text{school}}\}. \tag{5.10}$$

Vice versa, the trip time window can be propagated onto the school time window by (4.1), (4.3), and (4.10) or (4.16). Since school starts are required to be in discrete 5-minute time slots, rounding has an additional influence on the time windows. With $\lceil a \rceil_b := \lceil \frac{a}{b} \rceil \cdot b, \lfloor a \rfloor_b := \lfloor \frac{a}{b} \rfloor \cdot b$ for $a, b \in \mathbb{Q}, b \neq 0$ we obtain for all $(s,t) \in \mathcal{P}$

$$\begin{aligned}
\underline{\tau}_s &:= \max\{\underline{\tau}_s, \lceil \underline{\alpha}_t + \delta_{st}^{\text{school}} + \underline{\omega}_{st}^{\text{school}} \rceil_5\}, \\
\overline{\tau}_s &:= \min\{\overline{\tau}_s, \lfloor \overline{\alpha}_t + \delta_{st}^{\text{school}} + \overline{\omega}_{st}^{\text{school}} \rfloor_5\}.
\end{aligned} \tag{5.11}$$

The same idea can be applied for the time windows of feeder and collector trips, which are coupled by (4.9) or (4.15). Then the improved bounds for all $(t_1, t_2) \in \mathcal{C}$ are given by

$$\begin{aligned}
\overline{\alpha}_{t_1} &:= \min\{\overline{\alpha}_{t_1}, \overline{\alpha}_{t_2} + \delta_{t_1 t_2}^{\text{collector}} - \delta_{t_1 t_2}^{\text{feeder}} - \underline{\omega}_{t_1 t_2}^{\text{change}}\}, \\
\underline{\alpha}_{t_2} &:= \max\{\underline{\alpha}_{t_2}, \underline{\alpha}_{t_1} - \delta_{t_1 t_2}^{\text{collector}} + \delta_{t_1 t_2}^{\text{feeder}} + \underline{\omega}_{t_1 t_2}^{\text{change}}\}, \\
\overline{\alpha}_{t_2} &:= \min\{\overline{\alpha}_{t_2}, \overline{\alpha}_{t_1} - \delta_{t_1 t_2}^{\text{collector}} + \delta_{t_1 t_2}^{\text{feeder}} + \overline{\omega}_{t_1 t_2}^{\text{change}}\}, \\
\underline{\alpha}_{t_1} &:= \max\{\underline{\alpha}_{t_1}, \underline{\alpha}_{t_2} + \delta_{t_1 t_2}^{\text{collector}} - \delta_{t_1 t_2}^{\text{feeder}} - \overline{\omega}_{t_1 t_2}^{\text{change}}\}.
\end{aligned} \tag{5.12}$$

If some trips are connected in a schedule either by pre-existing deadhead trips (see Section 4.5) or if they are connected in later steps of the heuristic (see the next section) then we obtain further restriction for all $(t_1, t_2) \in \mathcal{A}$ with $x_{t_1 t_2} = 1$ from the constraints (4.8) or (4.14) on the trip time windows:

$$\begin{aligned}
\overline{\alpha}_{t_1} &:= \min\{\overline{\alpha}_{t_1}, \overline{\alpha}_{t_2} - \delta_{t_1}^{\text{trip}} - \delta_{t_1 t_2}^{\text{shift}} - \underline{\omega}_{t_1 t_2}^{\text{idle}}\}, \\
\underline{\alpha}_{t_2} &:= \max\{\underline{\alpha}_{t_2}, \underline{\alpha}_{t_1} + \delta_{t_1}^{\text{trip}} + \delta_{t_1 t_2}^{\text{shift}} + \underline{\omega}_{t_1 t_2}^{\text{idle}}\}, \\
\overline{\alpha}_{t_2} &:= \min\{\overline{\alpha}_{t_2}, \overline{\alpha}_{t_1} + \delta_{t_1}^{\text{trip}} + \delta_{t_1 t_2}^{\text{shift}} + \overline{\omega}_{t_1 t_2}^{\text{idle}}\}, \\
\underline{\alpha}_{t_1} &:= \max\{\underline{\alpha}_{t_1}, \underline{\alpha}_{t_2} - \delta_{t_1}^{\text{trip}} - \delta_{t_1 t_2}^{\text{shift}} - \overline{\omega}_{t_1 t_2}^{\text{idle}}\}.
\end{aligned} \tag{5.13}$$

In case there are cooperating schools we get for every $i \in \{1, \dots, C\}$ and for every $s_1, s_2 \in \mathcal{S}'_i, s_1 \leq s_2$ the following formulas from (4.29):

$$\begin{aligned}
\overline{\tau}_{s_1} &:= \min\{\overline{\tau}_{s_1}, \lfloor \hat{\tau}_{s_1} - \hat{\tau}_{s_2} + 5 \cdot \overline{\tau}_{s_2} \rfloor_5\}, \\
\underline{\tau}_{s_2} &:= \max\{\underline{\tau}_{s_2}, \lceil \hat{\tau}_{s_2} - \hat{\tau}_{s_1} + 5 \cdot \underline{\tau}_{s_1} \rceil_5\}, \\
\overline{\tau}_{s_2} &:= \min\{\overline{\tau}_{s_2}, \lfloor \hat{\tau}_{s_2} - \hat{\tau}_{s_1} + 5 \cdot \overline{\tau}_{s_1} \rfloor_5\}, \\
\underline{\tau}_{s_1} &:= \max\{\underline{\tau}_{s_1}, \lceil \hat{\tau}_{s_1} - \hat{\tau}_{s_2} + 5 \cdot \underline{\tau}_{s_2} \rceil_5\}.
\end{aligned} \tag{5.14}$$

In case there are trips in tacted lines we get for every $i \in \{1, \dots, T\}$ and for every $t_1, t_2 \in \mathcal{V}'_i, t_1 \leq t_2$ the following formulas from (4.30):

$$\begin{aligned}
\overline{\alpha}_{t_1} &:= \min\{\overline{\alpha}_{t_1}, \hat{\alpha}_{t_1} - \hat{\alpha}_{t_2} + \overline{\alpha}_{t_2}\}, \\
\underline{\alpha}_{t_2} &:= \max\{\underline{\alpha}_{t_2}, \hat{\alpha}_{t_2} - \hat{\alpha}_{t_1} + \underline{\alpha}_{t_1}\}, \\
\overline{\alpha}_{t_2} &:= \min\{\overline{\alpha}_{t_2}, \hat{\alpha}_{t_2} - \hat{\alpha}_{t_1} + \overline{\alpha}_{t_1}\}, \\
\underline{\alpha}_{t_1} &:= \max\{\underline{\alpha}_{t_1}, \hat{\alpha}_{t_1} - \hat{\alpha}_{t_2} + \underline{\alpha}_{t_2}\}.
\end{aligned} \tag{5.15}$$

Bounds strengthening steps (5.9) to (5.15) are iteratively repeated, until either no bound can be improved any more or an infeasibility occurs. That is, there is either a trip $t \in \mathcal{V}$

with $\underline{\alpha}_t > \overline{\alpha}_t$ or a school s with $\underline{\tau}_s > \overline{\tau}_s$ after applying one of the above update steps. An infeasibility means that there is no solution for the given instance due to a conflict of some time windows. In this case an IIS can be computed that can help the user to identify the inconsistencies in the input data. If the algorithm terminates without an infeasibility we have strengthened bounds on the α and τ variables.

In case of the multicriteria model these updated variable bounds are now used to calculate new bounds for the so far neglected ω variables. For all $(s,t) \in \mathcal{P}$ we have

$$
\begin{aligned}
\underline{\omega}_{st}^{\text{school}} &:= \max\{\underline{\omega}_{st}^{\text{school}}, 5 \cdot \underline{\tau}_s - \overline{\alpha}_t - \delta_{st}^{\text{school}}\}, \\
\overline{\omega}_{st}^{\text{school}} &:= \min\{\overline{\omega}_{st}^{\text{school}}, 5 \cdot \overline{\tau}_s - \underline{\alpha}_t - \delta_{st}^{\text{school}}\},
\end{aligned}
\tag{5.16}
$$

for all $(t_1, t_2) \in \mathcal{C}$ we have

$$
\begin{aligned}
\underline{\omega}_{t_1 t_2}^{\text{change}} &:= \max\{\underline{\omega}_{t_1 t_2}^{\text{change}}, \underline{\alpha}_{t_2} - \overline{\alpha}_{t_1} + \delta_{t_1 t_2}^{\text{collector}} - \delta_{t_1 t_2}^{\text{feeder}}\}, \\
\overline{\omega}_{t_1 t_2}^{\text{change}} &:= \min\{\overline{\omega}_{t_1 t_2}^{\text{change}}, \overline{\alpha}_{t_2} - \underline{\alpha}_{t_1} + \delta_{t_1 t_2}^{\text{collector}} - \delta_{t_1 t_2}^{\text{feeder}}\},
\end{aligned}
\tag{5.17}
$$

and for all $(t_1, t_2) \in \mathcal{A}$ with $x_{t_1 t_2} = 1$ we have

$$
\begin{aligned}
\underline{\omega}_{t_1 t_2}^{\text{idle}} &:= \max\{\underline{\omega}_{t_1 t_2}^{\text{idle}}, \underline{\alpha}_{t_2} - \overline{\alpha}_{t_1} - \delta_{t_1}^{\text{trip}} - \delta_{t_1 t_2}^{\text{shift}}\}, \\
\overline{\omega}_{t_1 t_2}^{\text{idle}} &:= \min\{\overline{\omega}_{t_1 t_2}^{\text{idle}}, \overline{\alpha}_{t_2} - \underline{\alpha}_{t_1} - \delta_{t_1}^{\text{trip}} - \delta_{t_1 t_2}^{\text{shift}}\}.
\end{aligned}
\tag{5.18}
$$

The iterated application of formulas (5.9) to (5.15) and thereafter the application of (5.16) to (5.18) (if necessary) is called *starting time propagation*. The starting time propagation is not only called once in the beginning to determine if the instance has a feasible solution at all but also several times in the course of the heuristic described below.

We finish this section with a (theoretical) estimation of the maximum (worst-case) runtime for the update of all bounds. If the bound strengthening from above is done in the same way as in the algorithm of Bar-Yehuda and Rawitz (Theorem 1) then it requires $O(mU)$ steps. Here m is the number of inequalities. Note that at most $|\mathcal{V}|$ many x-variables can be simultaneously set to their upper bound 1. Hence we get $m := 2 \cdot |\mathcal{P}| + 2 \cdot |\mathcal{C}| + 2 \cdot |\mathcal{V}|$ for the number of inequalities in our IP2. The variable U simply refers to the size of the largest time window, i.e., $U := \max\{U_1, U_2\}$ with $U_1 := \max\{\overline{\alpha}_t - \underline{\alpha}_t : t \in \mathcal{V}\}$ and $U_2 := \max\{\overline{\tau}_s - \underline{\tau}_s : s \in \mathcal{S}\}$.

Since every trip and every school has to start somewhen between midnight and midnight of the following day we already have $24 \cdot 60 = 1440$ as an upper bound on U. In the real-world instance we will consider later the value of U is even smaller, about $U = 300$ due to the concentration on the optimization of the morning peak. Thus the pseudo-polynomial starting time propagation algorithm for the bounds strengthening will turn out as a fast and powerful tool for practical purposes.

Constructing the Schedules

The construction of feasible schedules for all vehicles is the first stage of our heuristic, after the instance is checked to be feasible.

Consider the graph $(\mathcal{V}, \mathcal{A})$. In the k-th step of the heuristic a local-best deadhead-trip from $\mathcal{A}$ is selected. Before the actual selection (the evaluation of the scoring function) we remove those arcs from $\mathcal{A}$ that would lead to infeasibilities when being chosen. An arc $(t_1, t_2) \in \mathcal{A}$ is infeasible in view of inequalities (4.8) or (4.14) if either the vehicle arrives at the first bus stop of t_2 after the latest possible start of t_2 or it arrives so early that the maximum possible idle time limit is exceeded. Another kind of infeasibility occurs in view of inequalities (4.6) and (4.7). If some $x_{t_1 t_2}$ is already fixed to its upper bound 1 then all arcs $(t, t_2), (t_1, t) \in \mathcal{A}$ can also be removed from $\mathcal{A}$ since every tour must have a unique predecessor and successor, respectively. The same applies for trips that are already selected as first or last trip in some schedule.

In order to remove all those arcs that would make the solution infeasible when being selected we define $\mathcal{A}_k \subseteq \mathcal{A}$ as

$$
\begin{aligned}
\mathcal{A}_k := \mathcal{A}_{k-1} \backslash (\quad & \{(t_1, t_2) \in \mathcal{A} : \underline{\alpha}_{t_1} + \delta_{t_1}^{\text{trip}} + \delta_{t_1 t_2}^{\text{shift}} + \underline{\omega}_{t_1 t_2}^{\text{idle}} > \overline{\alpha}_{t_2}\} \\
\cup \ & \{(t_1, t_2) \in \mathcal{A} : \overline{\alpha}_{t_1} + \delta_{t_1}^{\text{trip}} + \delta_{t_1 t_2}^{\text{shift}} + \overline{\omega}_{t_1 t_2}^{\text{idle}} < \underline{\alpha}_{t_2}\} \\
\cup \ & \{(t_1, t_2) \in \mathcal{A} : \exists t \in \mathcal{V}, x_{t_1 t} = 1\} \\
\cup \ & \{(t_1, t_2) \in \mathcal{A} : \exists t \in \mathcal{V}, x_{t t_2} = 1\} \\
\cup \ & \{(t_1, t_2) \in \mathcal{A} : w_{t_1} = 1\} \\
\cup \ & \{(t_1, t_2) \in \mathcal{A} : v_{t_2} = 1\} \quad)
\end{aligned} \tag{5.19}
$$

for all $k \geq 1$ and $\mathcal{A}_0 := \mathcal{A}$. We set $x_{t_1 t_2} := 0$ for all removed arcs $(t_1, t_2) \in \mathcal{A}_{k-1} \backslash \mathcal{A}_k$.

As described in the previous section the scoring function is defined as

$$
s_{t_1 t_2} := \delta_{t_1 t_2}^{\text{shift}}, \tag{5.20}
$$

hence an arc $(t_1^*, t_2^*) \in \mathcal{A}_k$ with minimal distance $\delta_{t_1 t_2}^{\text{shift}}$ is selected, i.e.,

$$
(t_1^*, t_2^*) = \operatorname{argmin}\{s_{t_1 t_2} : (t_1, t_2) \in \mathcal{A}_k\}. \tag{5.21}
$$

We now set $x_{t_1^* t_2^*} := 1$. Due to the connection of trip t_1^* with t_2^* the starting time windows of these two trips might change. This might have effects on some or all of the other starting time windows of trips and schools. To evaluate the effect of the connection we redo the starting time propagation (5.9) to (5.15). In most cases we end up with updated feasible bounds on the time windows. However, in some cases an infeasibility occurs. Why this might happen is explained by the following example.

In contrast to the classical traveling salesman problem with static time windows feasibility of a given (partial) schedule is a more difficult issue. It is important to note that some arcs that lead to infeasible schedules are not removed from $\mathcal{A}_k$. As an example, consider the following, very small scenario shown in Figure 5.4: Two connectable trips, $\mathcal{V} := \{t_1, t_2\}, \mathcal{A} := \{(t_1, t_2)\}$, one school, $\mathcal{S} := \{s\}$, both trips are school trips, $\mathcal{P} := \{(s, t_1), (s, t_2)\}$, and no changes, $\mathcal{C} := \emptyset$. For trip t_1, the overall service duration is $\delta_{t_1}^{\text{trip}} := 25$ and the school is reached $\delta_{st_1}^{\text{school}} := 20$ minutes after the start. For trip t_2, the corresponding values are $\delta_{t_2}^{\text{trip}} := 25$ and $\delta_{st_2}^{\text{school}} := 10$. We assume the minimal and maximal waiting times for pupils at school in the interval $(\underline{\omega}_{st_i}^{\text{school}}, \overline{\omega}_{st_i}^{\text{school}}) := (5, 10)$ for both trips, $i = 1, 2$. The starting time of school s is in the interval $(\underline{\tau}_s, \overline{\tau}_s) := (450, 510)$. The deadhead trip duration between t_1 and t_2 is $\delta_{t_1 t_2}^{\text{shift}} := 5$. By starting time propagation

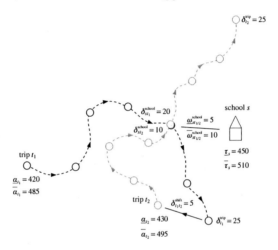

Figure 5.4: Connecting trip t_1 with t_2 is not feasible.

the school starting time is propagated onto both trip starting time windows. Thus, trip t_1 starts in the interval $(\underline{\alpha}_{t_1}, \overline{\alpha}_{t_1}) = (420, 485)$, and trip t_2 in $(\underline{\alpha}_{t_2}, \overline{\alpha}_{t_2}) = (430, 495)$. A bus serving t_1 and shifting to t_2, arrives at the start of t_2 within the interval $(450, 515)$. Since $\underline{\alpha}_{t_1} + \delta_{t_1}^{\text{trip}} + \delta_{t_1 t_2}^{\text{shift}} = 420 + 25 + 5 = 450 \leq \overline{\alpha}_{t_2} = 495$, the arc (t_1, t_2) is not removed by (5.19) and hence variable $x_{t_1 t_2}$ is not fixed to 0. However, setting $x_{t_1 t_2}$ to 1 is infeasible: It is easy to see that no $\alpha_{t_1}, \alpha_{t_2}, \tau_s$ can satisfy a system of inequalities with (4.14) for (t_1, t_2), and (4.10) for (s, t_1) and (s, t_2), that is:

$$
\begin{array}{rcll}
\alpha_{t_1} + 25 + 5 & \leq & \alpha_{t_2}, & \\
\alpha_{t_1} + 20 + 10 & \geq & 5 \cdot \tau_s, & (5.22) \\
\alpha_{t_2} + 10 + 5 & \leq & 5 \cdot \tau_s. &
\end{array}
$$

Detecting all these pitfalls can in principle be done by temporarily setting $x_{t_1 t_2}$ to 1 and checking feasibility of the resulting time windows for all $(t_1, t_2) \in \mathcal{A}$, one by one. However, this is very time-consuming. On the other hand, only a small fraction of all the deadheads in $\mathcal{A}_k$ are problematic, hence taking care of them only "on demand" turns out to be by far more efficient. For this, we copy the entire instance and the partial schedule generated so far, compute the local best shift, connect the corresponding trips and check feasibility by starting time propagation. If it turns out to be infeasible then we delete the corresponding arc in $\mathcal{A}_k$, i.e., $\mathcal{A}_k := \mathcal{A}_k \backslash \{(t_1^*, t_2^*)\}$, and search for the next local best shift.

The above steps are repeated iteratively until $\mathcal{A}_k = \emptyset$ for some k. This is equivalent to the fact that all binary variables x are fixed to their bounds, 0 or 1. Since v and w are slack variables their values are automatically determined by the x variables. Thus, we end up with a schedule for all buses and strengthened bounds on the time variables, which is now input for the second stage of the optimization.

---

schedules($\mathcal{I}$)

Input:	IOSANA instance $\mathcal{I}$
(1)	Call starting time propagation in $\mathcal{I}$
(2)	**If** starting time propagation returns "feasible" **Then**
(3)	Let $k := 0, \mathcal{A}_0 := \mathcal{A}$
(4)	**Repeat** the following steps
(5)	**Repeat** the following steps
(6)	Let $(t_1^*, t_2^*) = \text{argmin}\{s_{t_1 t_2} : (t_1, t_2) \in \mathcal{A}_k\}$
(7)	Copy instance $\mathcal{I}$ to $\mathcal{I}^*$
(8)	Set $x_{t_1^* t_2^*} = 1$ in $\mathcal{I}^*$
(9)	Call starting time propagation in $\mathcal{I}^*$
(10)	**If** starting time propagation returns "feasible" **Then**
(11)	Copy instance $\mathcal{I}^*$ to $\mathcal{I}$
(12)	Let $k := k + 1$
(13)	Let $\mathcal{A}_k := \mathcal{A}_{k-1} \backslash \{\ldots\}$ as in (5.19)
(14)	**Else**
(15)	Set $x_{t_1^* t_2^*} = 0$ in $\mathcal{I}$
(16)	Let $\mathcal{A}_k := \mathcal{A}_k \backslash \{(t_1^*, t_2^*)\}$
(17)	**End If**
(18)	**Until** starting time propagation returns "feasible" or $\mathcal{A}_k = \emptyset$
(19)	**Until** $\mathcal{A}_k = \emptyset$
(20)	Return feasible schedule x, v, w, new bounds on α, τ
(21)	**Else**
(22)	Return "problem infeasible"
(23)	**End If**
Output:	schedules v, w, x and new bounds on α, τ or message "problem infeasible"

Note that the above algorithm is not a greedy but an enumerative greedy heuristic since the selection of some deadhead trips might lead to infeasible partial solutions which has to be corrected when discovered. However, the discovery of an infeasibility happens right after the selection of a "wrong" deadhead trip so that the algorithm has to track back at most one single step.

From Greedy to PGreedy

The heuristic is so far an enumerative greedy heuristic that is able to produce feasible
solutions after a short amount of time, typically after a few seconds on a modern personal
computer. We extend the scoring function to a parametrized scoring function. If there
is more time available to the planner then a good set of parameters can be computed
using improving hit-and-run.

The scoring function (5.20) is by definition "blind" for everything that has to do with
time windows. However, in the VRPCTW and IOSANA the connection of two trips t_1
and t_2 does affect some or all of the time windows. And vice versa, the time windows
affect the set of deadhead trips that remain to be selected in the next round of the
heuristic. Thus, we seek a scoring function that does not only take into account the time
for the deadhead trip, but also takes care of the necessary changes in the corresponding
time windows. If the time windows are narrowed too early in the course of the algorithms,
the number of deployed buses quickly increases, because no "flexibility" remains. Thus,
we introduce a scoring function that prefers those connections that do not (or at least
not too much) change time windows of other trips or schools. For this, we define

$$
\begin{aligned}
s_{t_1 t_2}(\lambda) := \quad & \lambda_1 \cdot \delta_{t_1 t_2}^{\text{shift}} \\
+ \; & \lambda_2 \cdot |\underline{\alpha}_{t_1} + \delta_{t_1}^{\text{trip}} + \delta_{t_1 t_2}^{\text{shift}} + \underline{\omega}_{t_1 t_2}^{\text{idle}} - \underline{\alpha}_{t_2}| \\
+ \; & \lambda_3 \cdot |\overline{\alpha}_{t_1} + \delta_{t_1}^{\text{trip}} + \delta_{t_1 t_2}^{\text{shift}} + \underline{\omega}_{t_1 t_2}^{\text{idle}} - \overline{\alpha}_{t_2}| \\
+ \; & \lambda_4 \cdot |\underline{\alpha}_{t_1} + \delta_{t_1}^{\text{trip}} + \delta_{t_1 t_2}^{\text{shift}} + \underline{\omega}_{t_1 t_2}^{\text{idle}} - \overline{\alpha}_{t_2}| \\
+ \; & \lambda_5 \cdot |\overline{\alpha}_{t_1} + \delta_{t_1}^{\text{trip}} + \delta_{t_1 t_2}^{\text{shift}} + \underline{\omega}_{t_1 t_2}^{\text{idle}} - \underline{\alpha}_{t_2}|
\end{aligned}
\tag{5.23}
$$

with a fixed $\lambda = (\lambda_1, \ldots, \lambda_5) \in \mathbb{Q}_+^5$. (Note that this scoring function needs some additional terms if an upper bound on the maximum waiting time $\overline{\omega}_{t_1 t_2}^{\text{idle}}$ is specified.) Of
course, the length of the deadhead trip (the first addend) is again part of the score
$s_{t_1 t_2}(\lambda)$. The contribution of the second and the third addend to the score $s_{t_1 t_2}$ depends
on how good the time windows of trips t_1 and t_2 fit together; there is no contribution if
and only if the time windows perfectly coincide in the sense that the time windows for
trip t_1 at the first bus stop of trip t_2 (after deadheading) equals the starting time window
of trip t_2, i.e., $\underline{\alpha}_{t_1} + \delta_{t_1}^{\text{trip}} + \delta_{t_1 t_2}^{\text{shift}} = \underline{\alpha}_{t_2}$, and $\overline{\alpha}_{t_1} + \delta_{t_1}^{\text{trip}} + \delta_{t_1 t_2}^{\text{shift}} = \overline{\alpha}_{t_2}$. Finally, the contribution
of the last two addends in $s_{t_1 t_2}$ introduces a measure for the absolute differences between
the time windows of trips t_1 and t_2; there is no contribution if and only if trip t_1 arrives
at the first bus stop of t_2 (after deadheading) exactly in the moment when trip t_2 has
to start.

Since (5.23) is linear in the parameters Corollary 10 applies to our situation and we can
use formula (5.7). Hence we can apply improving hit-and-run for the search of a good set
of parameters. During our computational experiments we found out that negative values
for some λ_i always lead to poor solutions. Because of this observation it is reasonable to
restrict the domain of the parameters to the standard simplex $\{\lambda \in \mathbb{Q}_+^5 : \lambda_1 + \ldots + \lambda_5 =
1\}$, which narrows the search space even more.

Local Search Strategies

Local search heuristics, also called improvement heuristics, operate on a feasible solution of a combinatorial problem. In general, they try to improve the objective function value by performing local exchanges while maintaining feasibility. In principle, all local search techniques that were developed for the classical vehicle routing problem (see Laporte and Semet [51] for a survey) can be adapted for the VRPCTW and hence IOSANA. However, after each local exchange step we have to check whether the new schedules are still feasible with respect to the coupled time windows. Since the repeated checking of time windows is a time consuming process (remember that it involves the solution of an NP-complete problem by a pseudo-polynomial algorithm) we cannot consider exhaustive local search routines that involve the testing of many possible exchanges. We thus perform only the following simple local search step as a post-processing step after stage one is finished.

Assume (t_1, t_2) and (t_3, t_4) are connected by deadhead trips in a feasible solution of IOSANA, that is, $x_{t_1 t_2} = x_{t_3 t_4} = 1$ (see Figure 5.5). If $\delta^{\text{shift}}_{t_1 t_2} + \delta^{\text{shift}}_{t_3 t_4} > \delta^{\text{shift}}_{t_1 t_4} + \delta^{\text{shift}}_{t_3 t_2}$ we temporarily remove both deadhead trips and connect (t_1, t_4) and (t_3, t_2). That is, we let $x_{t_1 t_2} = x_{t_3 t_4} = 0$ and then set $x_{t_1 t_4} = x_{t_3 t_2} = 1$. Then we reset all time windows on α and τ to the starting values and redo the starting time propagation with respect to the new schedules, using once again the bounds strengthening steps (5.9) to (5.15). If the starting time propagation shows that the new schedules are also feasible then we accept the exchange. Otherwise, it is rejected.

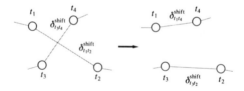

Figure 5.5: Local search.

This local search step is repeated for all pairs of deadhead trips, until no improvement is found any more. Note that this local search only reduces the total length of all deadhead trips, not the number of buses.

Assigning the Starting Times

The goal of stage two is the computation of school and trip starting times, given the bus schedules from stage one. During the computations in stage one, the time windows have been significantly narrowed. If a time window shrinks to a single point, then the corresponding variable, α_t or τ_s, can be fixed to this value. However, for the majority of all time windows, there is still some flexibility.

In this finishing step of the heuristic, starting times are assigned to each school and each bus trip. Given a schedule of the buses, the bounds on τ, α are tightened by starting time propagation.

times($\mathcal{I}, v, w, x$)	
Input:	IOSANA instance $\mathcal{I}$, schedules v, w, x
(1)	**For All** schools $s \in \mathcal{S}$ **Do**
(2)	**If** $\hat{\tau}_s \leq \underline{\tau}_s$ **Then**
(3)	Set $\tau_s = \frac{1}{5}\underline{\tau}_s$
(4)	**Else If** $\hat{\tau}_s \geq \overline{\tau}_s$ **Then**
(5)	Set $\tau_s = \frac{1}{5}\overline{\tau}_s$
(6)	**Else**
(7)	Set $\tau_s = \frac{1}{5}\hat{\tau}_s$
(8)	**End If**
(9)	Call starting time propagation
(10)	**End For**
(11)	**For All** bus trips $t \in \mathcal{V}$ **Do**
(12)	**If** $\hat{\alpha}_t \leq \underline{\alpha}_t$ **Then**
(13)	Set $\alpha_t = \underline{\alpha}_t$
(14)	**Else If** $\hat{\alpha}_t \geq \overline{\alpha}_t$ **Then**
(15)	Set $\alpha_t = \overline{\alpha}_t$
(16)	**Else**
(17)	Set $\alpha_t = \hat{\alpha}_t$
(18)	**End If**
(19)	Call starting time propagation
(20)	**End For**
Output:	starting times α and τ

We start with the schools and distinguish three cases. If the former school starting time $\hat{\tau}_s$ of school $s \in \mathcal{S}$ is below the lower bound $\underline{\tau}_s$, we fix the school starting time to the lower bound, that is, we let $\tau_s = \frac{1}{5}\underline{\tau}_s$, and the school s starts at $\underline{\tau}_s$. If the former school starting time of school s is above the upper bound $\overline{\tau}_s$, we fix the school starting time to the upper bound, that is, we let $\tau_s = \frac{1}{5}\overline{\tau}_s$, and the school s will start at $\overline{\tau}_s$. Finally, if the former school starting of school s is between the upper and the lower bound, $\underline{\tau}_s \leq \hat{\tau}_s \leq \overline{\tau}_s$, then this starting time is kept, i.e., we let $\tau_s = \frac{1}{5}\hat{\tau}_s$ and school s is going to start at $\hat{\tau}_s$. The idea behind this settings is to keep the new school starting times as close as possible to the former starting times that the pupils and teachers are already accustomed to. After each fixing of a school starting time, another round of starting time propagation has to be carried out to evaluate the impact of the respective setting for other school and trip starting times.

After having settled the school starting times, we assign starting times to the trips in the very same fashion. If the former trip starting time $\hat{\alpha}_t$ of bus trip $t \in \mathcal{V}$ is below the lower bound $\underline{\alpha}_t$, we fix the trip starting time to the lower bound, that is, we let $\alpha_t = \underline{\alpha}_t$.

If the former trip starting of trip t is above the upper bound $\overline{\alpha}_t$, then we fix the starting time to the upper bound, that is, we let $\alpha_t = \overline{\alpha}_t$. Finally, if the former trip starting of trip t is between the upper and the lower bound, $\underline{\alpha}_t \leq \hat{\alpha}_t \leq \overline{\alpha}_t$, then this starting time is kept, i.e., we let $\alpha_t = \hat{\alpha}_t$. Again, the starting times of the trips are settled in such way that the new starting times are as close as possible to the current ones. Also starting time propagation has to be called after the fixing of each trip starting time to evaluate the influences on the other trip starting times.

When the heuristic terminates after the second stage a feasible solution x, v, w, τ, α to the problem is returned.

Multicriteria Assignment of Starting Times

In stage two of the heuristic we presented above a simple way how to assign feasible starting times to the schools and the trips. The assignment itself was done in a greedy fashion. The schools are infinitely more important than the trips and in each of these two classes (schools and trips) the order of the assignment is following the order in which the instance is given (from first to last). For the bicriteria model this is of course sufficient, because there are no objective functions taking care of the different importance of the various starting and waiting times. However, in the multicriteria model we can do a bit better.

In contrast to stage one, where the objectives (4.11) and (4.12) are lexicographically ordered, there is no obvious ordering among the objectives (4.22), ..., (4.27) to solve the remaining multicriteria problem, i.e.,

$$
\begin{aligned}
\min \quad & ((4.22), \ldots, (4.27)) \\
\text{subject to} \quad & (4.14), \ldots, (4.21) \\
& (4.2), \ldots, (4.5) \\
& \tau \in \mathbb{Z}^{|\mathcal{S}|}, \alpha \in \mathbb{Z}^{|\mathcal{V}|} \\
& w^{\text{idle}} \in \mathbb{Z}^{|\mathcal{A}|}, w^{\text{school}} \in \mathbb{Z}^{|\mathcal{P}|}, w^{\text{change}} \in \mathbb{Z}^{|\mathcal{C}|} \\
& \Delta^v, \Delta^w \in \{0,1\}^{|\mathcal{V}|}, \Delta^x \in \{0,1\}^{|\mathcal{A}|} \\
& \Delta^\tau \in \mathbb{Z}^{|\mathcal{S}|}, \Delta^\alpha \in \mathbb{Z}^{|\mathcal{V}|}.
\end{aligned}
\tag{5.24}
$$

where v, w, x are the fixed variables corresponding to the bus schedule computed in stage one. Note that the trivial bounds on the variables (4.2), ..., (4.5) have changed due to the starting time propagations before and during stage one.

We can transform (5.24) into the a single-objective, parametric mixed-integer program by the weighted sum scalarization approach (see Section 2.2.4). A scalar vector $(\mu_1, \ldots, \mu_6) \geq 0, \mu_1 + \ldots + \mu_6 = 1$ is used to replace the multiple objectives by a single objective function. We then have to solve

$$
\begin{aligned}
\min \quad & \mu_1 \cdot (4.22) + \ldots + \mu_6 \cdot (4.27) \\
\text{subject to} \quad & (4.14), \ldots, (4.21) \\
& (4.2), \ldots, (4.5) \\
& \tau \in \mathbb{Z}^{|\mathcal{S}|}, \alpha \in \mathbb{Z}^{|\mathcal{V}|} \\
& \omega^{\text{idle}} \in \mathbb{Z}^{|\mathcal{A}|}, \omega^{\text{school}} \in \mathbb{Z}^{|\mathcal{P}|}, \omega^{\text{change}} \in \mathbb{Z}^{|\mathcal{C}|} \\
& \Delta^v, \Delta^w \in \{0,1\}^{|\mathcal{V}|}, \Delta^x \in \{0,1\}^{|\mathcal{A}|} \\
& \Delta^\tau \in \mathbb{Z}^{|\mathcal{S}|}, \Delta^\alpha \in \mathbb{Z}^{|\mathcal{V}|}.
\end{aligned}
\tag{5.25}
$$

Although this problem is theoretically difficult, for all real-world instances so far it could be solved to optimality by a standard MIP solver within a few seconds. However, it is necessary to select appropriate weights $\mu_1, \ldots, \mu_6$ that lead to good solutions. To support this selection process and make the possible alternatives visible for the planner we first examine the "size" of the solution space of stage two by computing lower and upper bounds $\underline{\varepsilon}_i, \overline{\varepsilon}_i$ for $i = 1, \ldots, 6$, one for each objective (4.22), ..., (4.27). For this, we solve a sequence of problems of type (5.25), where the negative resp. positive unit vectors are taken as weights in the objective function. That means, the minimization resp. maximization of only one goal is kept as objective function, the other goals are neglected. Thereafter, it is possible to compare each single objective value in the solution of (5.25) for a given set of weights $\mu_1, \ldots, \mu_6$ with its respective lower and upper bounds (see also Section 2.2.4). The weights are adjusted until an appropriate solution (with respect to the various political restrictions) is found, which then has to be implemented in real-world.

Chapter 6

Dual Aspects

Suppose we solve IOSANA using pgreedy as described in the previous chapter. Then for each new selection of λ parameters in the scoring function we potentially obtain another feasible solution with a different objective function value. The question is when to terminate this process. Of course the patience of the planner (i.e., the number of iterations) is limited. To evaluate the final solution we can compare it with all other solutions having a worse objective function value. Another way is to make use of dual information to obtain lower bounds. The comparison of upper bound (best feasible solution found by some heuristic) and lower bound (based on dual information) gives a measure for the quality of the solution. One way to obtain a lower bound is to use the LP relaxation. However, for real-world instances with large time windows this bound is sometimes of poor quality. In this chapter we present our efforts to improve the lower bound. We restrict our discussion to the bicriteria model (4.13), but the presented ideas can easily be adopted for the multicriteria model (4.28). In Section 6.1 we improve the LP relaxation of IOSANA by strengthening the inequalities and by adding further cutting planes. In Section 6.2 we discuss modeling alternatives that potentially lead to better bounds. Finally in Section 6.3 we show how primal and dual solution techniques can be combined to find even better solutions.

6.1 Improving the LP relaxation of IOSANA

The solutions generated by the pgreedy heuristic presented in the previous chapter look promising. They improve the current state by using much fewer buses. But of course, one never knows if better solutions, i.e., solutions with a fewer number of deployed buses, exist. This is, as a matter of fact, common to all heuristic approaches, no matter how sophisticated and smart they might be. The only chance to prove optimality is to improve the lower bound, in order to close the gap between lower and upper bound. Since IOSANA was formulated as a MIP a "natural" lower bound is given by its LP relaxation, that is, by solving instances of problem (4.13) while neglecting all integrality constraints. This can be done either only once or several times within an enumerative scheme such as branch-and-bound or branch-and-cut. In this section we discuss techniques that potentially improve the lower LP bound of (4.13).

6.1.1 Preprocessing

In general, *preprocessing* is the name for a bundle of techniques that are known to reduce the time for solving mixed integer programs (see Section 2.2.1).

Bounds Strengthening

For IOSANA it turned out that the starting time propagation (5.9) to (5.15) together with the fixing of binary variables thereafter (5.19) is not only a useful reduction of the search space for primal heuristics. It is moreover a powerful tool to improve the lower bound given by the LP relaxation. In general mixed-integer programming, our starting time propagation (see Section 5.2.1) is a special case of bounds strengthening, where bounds on all but one variable are used to obtain better bounds on the remaining variable (see Section 2.2.1). This procedure is included in the preprocessing or presolving routines in many commercial MIP-solver codes. However, only one inequality is used after the other to improve the bounds. Much stronger results can be obtained if more than one inequality is taken into consideration (two or three, for example). We now present the idea in general settings and discuss afterward its application to IOSANA.

Consider a mixed-integer program of the form (2.1). Assume there are two inequalities $a_{i_1}^T x \leq b_{i_1}, a_{i_2}^T x \leq b_{i_2}$ and a variable x_j with $a_{i_1 j} a_{i_2 j} < 0$. Then we multiply the first of the two inequalities by $|a_{i_2 j}|$ and the second one by $|a_{i_1 j}|$. Adding these two yields a new surrogate inequality of the form $a'^T x \leq b'$ with $a' = |a_{i_2 j}| \cdot a_{i_1} + |a_{i_1 j}| \cdot a_{i_2}$ and $b' = |a_{i_2 j}| \cdot b_{i_1} + |a_{i_1 j}| \cdot b_{i_2}$. Note that $a'_j = 0$. Then we apply for all remaining variables the bounds strengthening procedure as in Section 2.2.1 to obtain new bounds. The additional gain is that the bounds of variable x_j are not taken into consideration. Thus we have the chance to improve the bounds on the other variables even if x_j has weak bounds.

This idea can be extended to the case of three (or more) inequalities, where we have the chance to remove two variables. Assume we have identified three inequalities $a_{i_1}^T x \leq b_{i_1}, a_{i_2}^T x \leq b_{i_2}, a_{i_3}^T x \leq b_{i_3}$ and two variables x_{j_1}, x_{j_2} with

$$
\begin{aligned}
d_1 &:= a_{i_2 j_2} \cdot a_{i_3 j_1} - a_{i_3 j_2} \cdot a_{i_2 j_1} \geq 0, \\
d_2 &:= a_{i_1 j_1} \cdot a_{i_3 j_2} - a_{i_1 j_2} \cdot a_{i_3 j_1} \geq 0, \\
d_3 &:= a_{i_1 j_2} \cdot a_{i_2 j_1} - a_{i_1 j_1} \cdot a_{i_2 j_2} \geq 0.
\end{aligned}
\tag{6.1}
$$

Then a surrogate inequality is defined as $a' := d_1 a_{i_1} + d_2 a_{i_2} + d_3 a_{i_3}$ and $b' := d_1 b_{i_1} +$

$d_2 b_{i_2} + d_3 b_{i_3}$. Note that by definition $a'_{j_1} = a'_{j_2} = 0$. Then the bounds strengthening is applied to all remaining variables to obtain new bounds.

However, the time for doing an extensive presolve in the described way would increase considerably, thus this is not done by default in currently available codes. Since we know the structure of IOSANA a presolve using two or even three inequalities at a time can be carried out comparatively fast.

Consider two trips $(t_1, t_2) \in \mathcal{A}$ with $(t_1, t_2) \in \mathcal{C}$. Here the relevant subsystem of (4.13) consists of the first inequality of (4.8) and the second inequality of (4.9). That is,

$$
\begin{aligned}
\alpha_{t_1} + \delta_{t_1}^{\text{trip}} + \delta_{t_1 t_2}^{\text{shift}} + \underline{\omega}_{t_1 t_2}^{\text{idle}} - M \cdot (1 - x_{t_1 t_2}) &\leq \alpha_{t_2}, \\
\alpha_{t_1} + \delta_{t_1 t_2}^{\text{feeder}} + \overline{\omega}_{t_1 t_2}^{\text{change}} &\geq \alpha_{t_2} + \delta_{t_1 t_2}^{\text{collector}}.
\end{aligned}
\tag{6.2}
$$

By adding up these two inequalities we can eliminate the variables $\alpha_{t_1}, \alpha_{t_2}$ and obtain

$$
\delta_{t_1}^{\text{trip}} + \delta_{t_1 t_2}^{\text{shift}} + \underline{\omega}_{t_1 t_2}^{\text{idle}} - \delta_{t_1 t_2}^{\text{feeder}} - \overline{\omega}_{t_1 t_2}^{\text{change}} + \delta_{t_1 t_2}^{\text{collector}} \leq M \cdot (1 - x_{t_1 t_2}).
\tag{6.3}
$$

Hence if $\delta_{t_1}^{\text{trip}} + \delta_{t_1 t_2}^{\text{shift}} + \underline{\omega}_{t_1 t_2}^{\text{idle}} - \delta_{t_1 t_2}^{\text{feeder}} - \overline{\omega}_{t_1 t_2}^{\text{change}} + \delta_{t_1 t_2}^{\text{collector}} > 0$ we set $x_{t_1 t_2} := 0$ and thus eliminate this decision variable.

Using the second inequality of (4.8) and the first inequality of (4.9) we obtain after similar computations that we can set $x_{t_1 t_2} := 0$ if $\delta_{t_1}^{\text{trip}} + \delta_{t_1 t_2}^{\text{shift}} + \overline{\omega}_{t_1 t_2}^{\text{idle}} - \delta_{t_1 t_2}^{\text{feeder}} - \underline{\omega}_{t_1 t_2}^{\text{change}} + \delta_{t_1 t_2}^{\text{collector}} < 0$.

We can do more. Consider two trips $(t_1, t_2) \in \mathcal{A}$ with $(s, t_1), (s, t_2) \in \mathcal{P}$ for some school $s \in \mathcal{S}$. The relevant subsystem of (4.13) consists of the first inequality of (4.8), the second inequality of (4.10) for (s, t_1), and the first inequality of (4.10) for (s, t_2). That is,

$$
\begin{aligned}
\alpha_{t_1} + \delta_{t_1}^{\text{trip}} + \delta_{t_1 t_2}^{\text{shift}} + \underline{\omega}_{t_1 t_2}^{\text{idle}} - M \cdot (1 - x_{t_1 t_2}) &\leq \alpha_{t_2}, \\
\alpha_{t_1} + \delta_{s t_1}^{\text{school}} + \overline{\omega}_{s t_1}^{\text{school}} &\geq 5 \cdot \tau_s, \\
\alpha_{t_2} + \delta_{s t_2}^{\text{school}} + \underline{\omega}_{s t_2}^{\text{school}} &\leq 5 \cdot \tau_s.
\end{aligned}
\tag{6.4}
$$

By adding up these three inequalities we can eliminate the variables $\alpha_{t_1}, \alpha_{t_2}, \tau_s$ and obtain

$$
\delta_{t_1}^{\text{trip}} + \delta_{t_1 t_2}^{\text{shift}} + \underline{\omega}_{t_1 t_2}^{\text{idle}} - \delta_{s t_1}^{\text{school}} - \overline{\omega}_{s t_1}^{\text{school}} + \delta_{s t_2}^{\text{school}} + \underline{\omega}_{s t_2}^{\text{school}} \leq M \cdot (1 - x_{t_1 t_2}).
\tag{6.5}
$$

Thus if $\delta_{t_1}^{\text{trip}} + \delta_{t_1 t_2}^{\text{shift}} + \underline{\omega}_{t_1 t_2}^{\text{idle}} - \delta_{s t_1}^{\text{school}} - \overline{\omega}_{s t_1}^{\text{school}} + \delta_{s t_2}^{\text{school}} + \underline{\omega}_{s t_2}^{\text{school}} > 0$ we set $x_{t_1 t_2} := 0$.

Note that this was already described in the primal greedy heuristic in Section 5.2.1. In the heuristic however it was only necessary to deal with this problem on demand, i.e., when an infeasible $x_{t_1 t_2}$ is selected. Most of them are never chosen anyway. In the dual approach the situation is different. Here it pays to remove these infeasible variables to obtain better bounds and reduce the size of the branch-and-bound tree.

Using the second inequality of (4.8), the first inequality of (4.10) for (s, t_1), and the second inequality of (4.10) for (s, t_2) we obtain that we can set $x_{t_1 t_2} := 0$ if

$$
\delta_{t_1}^{\text{trip}} + \delta_{t_1 t_2}^{\text{shift}} + \overline{\omega}_{t_1 t_2}^{\text{idle}} - \delta_{s t_1}^{\text{school}} - \underline{\omega}_{s t_1}^{\text{school}} + \delta_{s t_2}^{\text{school}} + \overline{\omega}_{s t_2}^{\text{school}} < 0.
\tag{6.6}
$$

Big-M Reduction

Consider the inequalities in (4.8). Informally speaking, due to the "big-M"-formulation they have the disadvantage of not being very strong, that is, even when they are removed from the model, the objective function value of the LP relaxation does not change much. The x and the α variables are only loosely linked in the LP relaxation. Moreover, it is known from experience that models having "big-M"-inequalities tend to cause computational problems. One way to reduce (but not eliminate) the problem is to compute best possible values for M so that the corresponding inequalities become as strong as possible.

Let $(t_1, t_2) \in \mathcal{A}$. It was already noted that M depends on (t_1, t_2). We strengthen these inequalities by computing the best possible value for $M_{t_1 t_2}$. The first inequality in (4.8) is weak for $x_{t_1 t_2} = 0$ in the sense that for all $\alpha_{t_1} \in [\underline{\alpha}_{t_1}, \overline{\alpha}_{t_1}]$ and all $\alpha_{t_2} \in [\underline{\alpha}_{t_2}, \overline{\alpha}_{t_2}]$ we have

$$\alpha_{t_1} + \delta_{t_1}^{\text{trip}} + \delta_{t_1 t_2}^{\text{shift}} + \underline{\omega}_{t_1 t_2}^{\text{idle}} - M'_{t_1 t_2} \le \alpha_{t_2}. \tag{6.7}$$

Therefore we get

$$\overline{\alpha}_{t_1} + \delta_{t_1}^{\text{trip}} + \delta_{t_1 t_2}^{\text{shift}} + \underline{\omega}_{t_1 t_2}^{\text{idle}} - M'_{t_1 t_2} \le \underline{\alpha}_{t_2}, \tag{6.8}$$

thus we can set

$$M'_{t_1 t_2} := \overline{\alpha}_{t_1} - \underline{\alpha}_{t_2} + \delta_{t_1}^{\text{trip}} + \delta_{t_1 t_2}^{\text{shift}} + \underline{\omega}_{t_1 t_2}^{\text{idle}}. \tag{6.9}$$

For the second inequality in (4.8) we obtain by similar computations

$$M''_{t_1 t_2} := \overline{\alpha}_{t_2} - \underline{\alpha}_{t_1} - \delta_{t_1}^{\text{trip}} - \delta_{t_1 t_2}^{\text{shift}} - \overline{\omega}_{t_1 t_2}^{\text{idle}}. \tag{6.10}$$

Note that these "big-M"-computations are carried out after the starting times propagation, because there the bounds on α might be changed which affects (improves) also the values for $M'_{t_1 t_2}$ and $M''_{t_1 t_2}$, respectively.

Lifting Coefficients

Now inequality (4.8) can be further strengthened by lifting an additional binary variable. For this, we apply a lifting technique developed by Desrochers and Laporte [27] (see also Kara, Laporte, and Bektas [45]) for the vehicle routing problem with time windows, VRPTW. (More general, lifting as a technique for general integer programming to strengthen valid inequalities was introduced by Padberg [61] and Wolsey [76].) Their new inequality is based on the observation that $x_{t_1 t_2} + x_{t_2 t_1} \le 1$ for all $(t_1, t_2) \in \mathcal{A}$ with also $(t_2, t_1) \in \mathcal{A}$.

Theorem 11 *For all $(t_1, t_2) \in \mathcal{A}$ with $(t_2, t_1) \in \mathcal{A}$ the constraints*

$$\alpha_{t_1} + \delta_{t_1}^{\text{trip}} + \delta_{t_1 t_2}^{\text{shift}} + \underline{\omega}_{t_1 t_2}^{\text{idle}} \quad - (\overline{\alpha}_{t_1} - \underline{\alpha}_{t_2} + \delta_{t_1}^{\text{trip}} + \delta_{t_1 t_2}^{\text{shift}} + \underline{\omega}_{t_1 t_2}^{\text{idle}}) \cdot (1 - x_{t_1 t_2})$$
$$+ \max\{0, \overline{\alpha}_{t_1} - \underline{\alpha}_{t_2} - \delta_{t_2}^{\text{trip}} - \delta_{t_2 t_1}^{\text{shift}} - \overline{\omega}_{t_2 t_1}^{\text{idle}}\} \cdot x_{t_2 t_1} \le \alpha_{t_2} \tag{6.11}$$

and

$$\alpha_{t_1} + \delta_{t_1}^{\text{trip}} + \delta_{t_1 t_2}^{\text{shift}} + \overline{\omega}_{t_1 t_2}^{\text{idle}} \quad + (\overline{\alpha}_{t_2} - \underline{\alpha}_{t_1} + \delta_{t_1}^{\text{trip}} + \delta_{t_1 t_2}^{\text{shift}} + \overline{\omega}_{t_1 t_2}^{\text{idle}}) \cdot (1 - x_{t_1 t_2})$$
$$- \max\{0, \overline{\alpha}_{t_2} - \underline{\alpha}_{t_1} - \delta_{t_2}^{\text{trip}} - \delta_{t_2 t_1}^{\text{shift}} - \underline{\omega}_{t_2 t_1}^{\text{idle}}\} \cdot x_{t_2 t_1} \geq \alpha_{t_2}$$

$$(6.12)$$

are valid inequalities for (4.13).

Proof. Consider for an arbitrary $(t_1, t_2) \in \mathcal{A}$ with $(t_2, t_1) \in \mathcal{A}$ inequality (4.8) with $M'_{t_1 t_2}$ as "big-M", that is

$$\alpha_{t_1} + \delta_{t_1}^{\text{trip}} + \delta_{t_1 t_2}^{\text{shift}} + \underline{\omega}_{t_1 t_2}^{\text{idle}} - (\overline{\alpha}_{t_1} - \underline{\alpha}_{t_2} + \delta_{t_1}^{\text{trip}} + \delta_{t_1 t_2}^{\text{shift}} + \underline{\omega}_{t_1 t_2}^{\text{idle}}) \cdot (1 - x_{t_1 t_2}) \leq \alpha_{t_2}. \quad (6.13)$$

We introduce an additional term $\xi_{t_2 t_1} \cdot x_{t_2 t_1}$ in (6.13) and obtain

$$\alpha_{t_1} + \delta_{t_1}^{\text{trip}} + \delta_{t_1 t_2}^{\text{shift}} + \underline{\omega}_{t_1 t_2}^{\text{idle}} - (\overline{\alpha}_{t_1} - \underline{\alpha}_{t_2} + \delta_{t_1}^{\text{trip}} + \delta_{t_1 t_2}^{\text{shift}} + \underline{\omega}_{t_1 t_2}^{\text{idle}}) \cdot (1 - x_{t_1 t_2}) + \xi_{t_2 t_1} \cdot x_{t_2 t_1} \leq \alpha_{t_2}. \quad (6.14)$$

We are seeking the "best" (i.e., largest) value for $\xi_{t_2 t_1}$. The new term only occurs when $x_{t_2 t_1} = 1$. In this case we have $x_{t_1 t_2} = 0$. Thus (6.14) reads:

$$\alpha_{t_1} + \delta_{t_1}^{\text{trip}} + \delta_{t_1 t_2}^{\text{shift}} + \underline{\omega}_{t_1 t_2}^{\text{idle}} - (\overline{\alpha}_{t_1} - \underline{\alpha}_{t_2} + \delta_{t_1}^{\text{trip}} + \delta_{t_1 t_2}^{\text{shift}} + \underline{\omega}_{t_1 t_2}^{\text{idle}}) + \xi_{t_2 t_1} \leq \alpha_{t_2}, \quad (6.15)$$

hence $\xi_{t_2 t_1}$ must be chosen in such way that

$$\xi_{t_2 t_1} \leq \alpha_{t_2} - \alpha_{t_1} + \overline{\alpha}_{t_1} - \underline{\alpha}_{t_2} \quad (6.16)$$

is fulfilled. From inequality (4.8) for (t_2, t_1) we obtain with $x_{t_2 t_1} = 1$:

$$\alpha_{t_2} + \delta_{t_2}^{\text{trip}} + \delta_{t_2 t_1}^{\text{shift}} + \overline{\omega}_{t_2 t_1}^{\text{idle}} + M \cdot (1 - 1) \geq \alpha_{t_1}, \quad (6.17)$$

and thus

$$\alpha_{t_2} - \alpha_{t_1} \geq -\delta_{t_2}^{\text{trip}} - \delta_{t_2 t_1}^{\text{shift}} - \overline{\omega}_{t_2 t_1}^{\text{idle}}. \quad (6.18)$$

So we can estimate

$$\overline{\alpha}_{t_1} - \underline{\alpha}_{t_2} - \delta_{t_2}^{\text{trip}} - \delta_{t_2 t_1}^{\text{shift}} - \overline{\omega}_{t_2 t_1}^{\text{idle}} \leq \alpha_{t_2} - \alpha_{t_1} + \overline{\alpha}_{t_1} - \underline{\alpha}_{t_2}. \quad (6.19)$$

Now we set

$$\xi_{t_2 t_1} := \overline{\alpha}_{t_1} - \underline{\alpha}_{t_2} - \delta_{t_2}^{\text{trip}} - \delta_{t_2 t_1}^{\text{shift}} - \overline{\omega}_{t_2 t_1}^{\text{idle}}, \quad (6.20)$$

which yields the proposition. (Note that for $\xi_{t_2 t_1} \leq 0$, the coefficient does not need to be lifted into the inequality.)

The proof for inequality (6.12) is similar, thus we skip the details. $\square$

Replacing (4.8) by (6.11) and (6.12), respectively, leads to a tighter LP relaxation of the model.

The next theorem is also inspired by the article of Desrochers and Laporte [27]. Here we take the trivial inequalities $\underline{\alpha} \leq \alpha \leq \overline{\alpha}$ and introduce x-variables by lifting.

Theorem 12 *For all $t_1 \in \mathcal{V}$ the constraints*

$$\underline{\alpha}_{t_1} + \sum_{t_0:(t_0,t_1)\in\mathcal{A}} \max\{0, \underline{\alpha}_{t_0} - \underline{\alpha}_{t_1} + \delta_{t_0}^{\text{trip}} + \delta_{t_0t_1}^{\text{shift}} + \underline{\omega}_{t_0t_1}^{\text{idle}}\} \cdot x_{t_0t_1} \leq \underline{\alpha}_{t_1}, \quad (6.21)$$

$$\overline{\alpha}_{t_1} - \sum_{t_0:(t_0,t_1)\in\mathcal{A}} \max\{0, \underline{\alpha}_{t_1} - \underline{\alpha}_{t_0} - \delta_{t_0}^{\text{trip}} - \delta_{t_0t_1}^{\text{shift}} - \overline{\omega}_{t_0t_1}^{\text{idle}}\} \cdot x_{t_0t_1} \geq \overline{\alpha}_{t_1}, \quad (6.22)$$

$$\overline{\alpha}_{t_1} - \sum_{t_2:(t_1,t_2)\in\mathcal{A}} \max\{0, \overline{\alpha}_{t_1} - \overline{\alpha}_{t_2} + \delta_{t_1}^{\text{trip}} + \delta_{t_1t_2}^{\text{shift}} + \underline{\omega}_{t_1t_2}^{\text{idle}}\} \cdot x_{t_1t_2} \geq \overline{\alpha}_{t_1}, \quad (6.23)$$

$$\underline{\alpha}_{t_1} + \sum_{t_2:(t_1,t_2)\in\mathcal{A}} \max\{0, \underline{\alpha}_{t_2} - \underline{\alpha}_{t_1} - \delta_{t_1}^{\text{trip}} - \delta_{t_1t_2}^{\text{shift}} - \overline{\omega}_{t_1t_2}^{\text{idle}}\} \cdot x_{t_1t_2} \leq \underline{\alpha}_{t_1} \quad (6.24)$$

are valid inequalities for (4.13).

Proof. To prove inequality (6.21), let $t_2 \in \mathcal{V}$. By the lower bound inequality of (4.1) on the trip starting time we have $\underline{\alpha}_{t_2} \leq \alpha_{t_2}$. For each $t_1 \in \mathcal{V}$ with $(t_1, t_2) \in \mathcal{A}$ we introduce an additional term $\xi_{t_1t_2} \cdot x_{t_1t_2}$ and compute the largest possible value for $\xi_{t_1t_2}$. Note that these computations can be done sequentially, because for at most one t_1 we have $x_{t_1t_2} = 1$, thus all $\xi_{t_1t_2}$ are independent of each other. For $x_{t_1t_2} = 0$ we can assume an arbitrary value for $\xi_{t_1t_2}$. For $x_{t_1t_2} = 1$ we obtain from the first inequality of (4.8)

$$\begin{aligned} \alpha_{t_2} &\geq \alpha_{t_1} + \delta_{t_1}^{\text{trip}} + \delta_{t_1t_2}^{\text{shift}} + \omega_{t_1t_2}^{\text{idle}} \\ &\geq \underline{\alpha}_{t_1} + \delta_{t_1}^{\text{trip}} + \delta_{t_1t_2}^{\text{shift}} + \underline{\omega}_{t_1t_2}^{\text{idle}} \\ &= \underline{\alpha}_{t_2} + (-\underline{\alpha}_{t_2} + \underline{\alpha}_{t_1} + \delta_{t_1}^{\text{trip}} + \delta_{t_1t_2}^{\text{shift}} + \underline{\omega}_{t_1t_2}^{\text{idle}}). \end{aligned} \quad (6.25)$$

If $-\underline{\alpha}_{t_2} + \underline{\alpha}_{t_1} + \delta_{t_1}^{\text{trip}} + \delta_{t_1t_2}^{\text{shift}} + \underline{\omega}_{t_1t_2}^{\text{idle}}$ is negative then 0 is a better bound because of (4.1). Thus we obtain $\xi_{t_1t_2} := \max\{0, -\underline{\alpha}_{t_2} + \underline{\alpha}_{t_1} + \delta_{t_1}^{\text{trip}} + \delta_{t_1t_2}^{\text{shift}} + \underline{\omega}_{t_1t_2}^{\text{idle}}\}$.

The proof of the other inequalities (6.22), (6.23), and (6.24) can be carried out analogously, so we omit the details.   □

We use the techniques from the previous theorem to lift coefficients in the trivial inequalities for the τ variables.

Theorem 13 *For all $(s, t_1) \in \mathcal{P}$ the constraints*

$$\frac{1}{5}\overline{\tau}_s \cdot w_{t_1} + \sum_{t_2:(t_1,t_2)\in\mathcal{A}} \lfloor \overline{\alpha}_{t_2} - \delta_{t_1}^{\text{trip}} - \delta_{t_1t_2}^{\text{shift}} - \omega_{t_1t_2}^{\text{idle}} + \delta_{st_1}^{\text{school}} + \overline{\omega}_{st_1}^{\text{school}}\rfloor_5 \cdot x_{t_1t_2} \geq \tau_s \quad (6.26)$$

$$\frac{1}{5}\overline{\tau}_s \cdot v_{t_1} + \sum_{t_0:(t_0,t_1)\in\mathcal{A}} \lfloor \overline{\alpha}_{t_0} + \delta_{t_0}^{\text{trip}} + \delta_{t_0t_1}^{\text{shift}} + \overline{\omega}_{t_0t_1}^{\text{idle}} + \delta_{st_1}^{\text{school}} + \overline{\omega}_{st_1}^{\text{school}}\rfloor_5 \cdot x_{t_0t_1} \geq \tau_s \quad (6.27)$$

$$\frac{1}{5}\underline{\tau}_s \cdot w_{t_1} + \sum_{t_2:(t_1,t_2)\in\mathcal{A}} \lceil \underline{\alpha}_{t_2} - \delta_{t_1}^{\text{trip}} - \delta_{t_1t_2}^{\text{shift}} - \overline{\omega}_{t_1t_2}^{\text{idle}} + \delta_{st_1}^{\text{school}} + \underline{\omega}_{st_1}^{\text{school}}\rceil_5 \cdot x_{t_1t_2} \leq \tau_s \quad (6.28)$$

$$\frac{1}{5}\underline{\tau}_s \cdot v_{t_1} + \sum_{t_0:(t_0,t_1)\in\mathcal{A}} \lceil \underline{\alpha}_{t_0} + \delta_{t_0}^{\text{trip}} + \delta_{t_0t_1}^{\text{shift}} + \underline{\omega}_{t_0t_1}^{\text{idle}} + \delta_{st_1}^{\text{school}} + \underline{\omega}_{st_1}^{\text{school}}\rceil_5 \cdot x_{t_0t_1} \leq \tau_s \quad (6.29)$$

are valid inequalities for (4.13).

Proof. We only prove (6.26) since the other three inequalities can be shown similarly. Let $s \in \mathcal{S}$ be an arbitrary school and $t_1 \in \mathcal{V}$ a trip with $(s, t_1) \in \mathcal{P}$. Because of (4.7) there is at most one trip $t_2 \in \mathcal{V}$ with $(t_1, t_2) \in \mathcal{A}$ and $x_{t_1 t_2} = 1$. Using inequalities (4.8) and (4.10) we can estimate the upper bounds on the school starting time in this case as

$$5\tau_s \leq \alpha_{t_1} + \delta_{st_1}^{\text{school}} + \overline{\omega}_{st_1}^{\text{school}} \leq \overline{\alpha}_{t_2} - \delta_{t_1}^{\text{trip}} - \delta_{t_1 t_2}^{\text{shift}} - \underline{\omega}_{t_1 t_2}^{\text{idle}} + \delta_{st_1}^{\text{school}} + \overline{\omega}_{st_1}^{\text{school}}. \tag{6.30}$$

After dividing this inequality by 5 we can round down the right-hand side. That is,

$$\tau_s \leq \lfloor \overline{\alpha}_{t_2} - \delta_{t_1}^{\text{trip}} - \delta_{t_1 t_2}^{\text{shift}} - \underline{\omega}_{t_1 t_2}^{\text{idle}} + \delta_{st_1}^{\text{school}} + \overline{\omega}_{st_1}^{\text{school}} \rfloor_5. \tag{6.31}$$

If there is no such trip t_2 then $w_{t_1} = 1$ because of (4.7). In this case we have $\tau_s \leq \frac{1}{5}\overline{\tau}_s$. Putting this together we obtain inequality (6.26). □

Valid Inequalities

The following result can be seen as an application of the general Chvatal-Gomory procedure to construct a valid inequality. Inequality (6.32) is thus called *CG school cut* in the sequel.

Theorem 14 *For all* $(s_1, t_1), (s_2, t_2) \in \mathcal{P}$ *with* $(t_1, t_2) \in \mathcal{A}$ *the constraints*

$$\tau_{s_1} - \tau_{s_2} \leq \lfloor \delta_{s_1 t_1}^{\text{school}} + \overline{\omega}_{s_1 t_1}^{\text{school}} - \delta_{t_1}^{\text{trip}} - \delta_{t_1 t_2}^{\text{shift}} - \underline{\omega}_{t_1 t_2}^{\text{idle}} - \delta_{s_2 t_2}^{\text{school}} - \underline{\omega}_{s_2 t_2}^{\text{school}} \rfloor_5 \\ + M \cdot (1 - x_{t_1 t_2}) \tag{6.32}$$

with

$$M := \min\{\overline{\tau}_{s_1} - \underline{\tau}_{s_2} - \lfloor \delta_{s_1 t_1}^{\text{school}} + \overline{\omega}_{s_1 t_1}^{\text{school}} - \delta_{t_1}^{\text{trip}} - \delta_{t_1 t_2}^{\text{shift}} - \underline{\omega}_{t_1 t_2}^{\text{idle}} - \delta_{s_2 t_2}^{\text{school}} - \underline{\omega}_{s_2 t_2}^{\text{school}} \rfloor_5, \\ (\lfloor \overline{\alpha}_{t_1} - \underline{\alpha}_{t_2} + \delta_{t_1}^{\text{trip}} + \delta_{t_1 t_2}^{\text{shift}} + \underline{\omega}_{t_1 t_2}^{\text{idle}} \rfloor_5 + 1)\} \tag{6.33}$$

and

$$\tau_{s_2} - \tau_{s_1} \leq \lfloor \delta_{s_2 t_2}^{\text{school}} + \overline{\omega}_{s_2 t_2}^{\text{school}} + \delta_{t_1}^{\text{trip}} + \delta_{t_1 t_2}^{\text{shift}} + \overline{\omega}_{t_1 t_2}^{\text{idle}} - \delta_{s_1 t_1}^{\text{school}} - \underline{\omega}_{s_1 t_1}^{\text{school}} \rfloor_5 \\ + M \cdot (1 - x_{t_1 t_2}) \tag{6.34}$$

with

$$M := \min\{\overline{\tau}_{s_2} - \underline{\tau}_{s_1} - \lfloor \delta_{s_2 t_2}^{\text{school}} + \overline{\omega}_{s_2 t_2}^{\text{school}} + \delta_{t_1}^{\text{trip}} + \delta_{t_1 t_2}^{\text{shift}} + \overline{\omega}_{t_1 t_2}^{\text{idle}} - \delta_{s_1 t_1}^{\text{school}} - \underline{\omega}_{s_1 t_1}^{\text{school}} \rfloor_5, \\ (\lfloor \overline{\alpha}_{t_2} - \underline{\alpha}_{t_1} - \delta_{t_1}^{\text{trip}} - \delta_{t_1 t_2}^{\text{shift}} - \overline{\omega}_{t_1 t_2}^{\text{idle}} \rfloor_5 + 1)\} \tag{6.35}$$

are valid inequalities for (4.13).

Proof. To prove inequality (6.32) consider for some $(s_1, t_1), (s_2, t_2) \in \mathcal{P}$ with $(t_1, t_2) \in \mathcal{A}$ the model constraints

$$\begin{aligned} \alpha_{t_1} + \delta_{t_1}^{\text{trip}} + \delta_{t_1 t_2}^{\text{shift}} + \underline{\omega}_{t_1 t_2}^{\text{idle}} &\leq \alpha_{t_2} + (\overline{\alpha}_{t_1} - \underline{\alpha}_{t_2} + \delta_{t_1}^{\text{trip}} + \delta_{t_1 t_2}^{\text{shift}} + \underline{\omega}_{t_1 t_2}^{\text{idle}}) \cdot (1 - x_{t_1 t_2}), \\ \alpha_{t_1} + \delta_{s_1 t_1}^{\text{school}} + \overline{\omega}_{s_1 t_1}^{\text{school}} &\geq 5 \cdot \tau_{s_1}, \\ \alpha_{t_2} + \delta_{s_2 t_2}^{\text{school}} + \underline{\omega}_{s_2 t_2}^{\text{school}} &\leq 5 \cdot \tau_{s_2}. \end{aligned} \tag{6.36}$$

By adding up these three inequalities we obtain

$$
\begin{aligned}
5 \cdot \tau_{s_1} - 5 \cdot \tau_{s_2} \leq\ & \delta^{\text{school}}_{s_1 t_1} + \overline{\omega}^{\text{school}}_{s_1 t_1} - \delta^{\text{trip}}_{t_1} - \delta^{\text{shift}}_{t_1 t_2} - \underline{\omega}^{\text{idle}}_{t_1 t_2} - \delta^{\text{school}}_{s_2 t_2} - \underline{\omega}^{\text{school}}_{s_2 t_2} \\
& + \left(\overline{\alpha}_{t_1} - \underline{\alpha}_{t_2} + \delta^{\text{trip}}_{t_1} + \delta^{\text{shift}}_{t_1 t_2} + \underline{\omega}^{\text{idle}}_{t_1 t_2} \right) \cdot \left(1 - x_{t_1 t_2} \right).
\end{aligned}
\tag{6.37}
$$

Now we can divide this new inequality by 5 and round down the right-hand side. With $\lfloor a + b \rfloor \leq \lfloor a \rfloor + \lfloor b \rfloor + 1$ we obtain

$$
\begin{aligned}
\tau_{s_1} - \tau_{s_2} \leq\ & \left\lfloor \delta^{\text{school}}_{s_1 t_1} + \overline{\omega}^{\text{school}}_{s_1 t_1} - \delta^{\text{trip}}_{t_1} - \delta^{\text{shift}}_{t_1 t_2} - \underline{\omega}^{\text{idle}}_{t_1 t_2} - \delta^{\text{school}}_{s_2 t_2} - \underline{\omega}^{\text{school}}_{s_2 t_2} \right\rfloor_5 \\
& + \left(\left\lfloor \overline{\alpha}_{t_1} - \underline{\alpha}_{t_2} + \delta^{\text{trip}}_{t_1} + \delta^{\text{shift}}_{t_1 t_2} + \underline{\omega}^{\text{idle}}_{t_1 t_2} \right\rfloor_5 + 1 \right) \cdot \left(1 - x_{t_1 t_2} \right).
\end{aligned}
\tag{6.38}
$$

A potentially better coefficient for $(1 - x_{t_1 t_2})$ can be obtained from the bounds on τ, that is

$$
\overline{\tau}_{s_1} - \underline{\tau}_{s_2} - \left\lfloor \delta^{\text{school}}_{s_1 t_1} + \overline{\omega}^{\text{school}}_{s_1 t_1} - \delta^{\text{trip}}_{t_1} - \delta^{\text{shift}}_{t_1 t_2} - \underline{\omega}^{\text{idle}}_{t_1 t_2} - \delta^{\text{school}}_{s_2 t_2} - \underline{\omega}^{\text{school}}_{s_2 t_2} \right\rfloor_5.
\tag{6.39}
$$

Taking the minimum of both in (6.33) we get inequality (6.32).

The proof of inequality (6.34) can be carried out similarly.   □

6.1.2   Cutting Planes

The cutting planes presented in this section are also valid inequalities. In contrast to the inequalities presented in Theorems 11, 12, and 13 the inequalities here are too many to be added in advance to the model. In this case the linear programming relaxation would become too big. In contrast we generate them only on demand within a branch-and-cut framework.

k-Path Cuts

In order to solve instances of the VRPTW Kohl et al. [48] (see also the survey of Cordeau et al. [22]) introduced the following cuts which we adapt to solve instances of IOSANA.

For a subset of the trips $\mathcal{W} \subseteq \mathcal{V}$ we define the following variable representing the flow into $\mathcal{W}$

$$
f(\mathcal{W}) := \sum_{t_1 \in \mathcal{V} \setminus \mathcal{W}} \sum_{t_2 \in \mathcal{W}} x_{t_1 t_2} + \sum_{t_2 \in \mathcal{W}} v_{t_2}.
\tag{6.40}
$$

Let $k(\mathcal{W})$ be the smallest number of buses necessary to serve all trips in $\mathcal{W}$. Then the constraint

$$
f(\mathcal{W}) \geq k(\mathcal{W})
\tag{6.41}
$$

is called a k-path cut. At least k paths (pull-in or deadhead trips) have to enter subset $\mathcal{W}$ in a feasible solution for IOSANA.

In order to use (6.41) as cutting planes to cut off infeasible fractional LP-optimal solutions we have to address two problems. Given an LP-optimal solution of IOSANA how to quickly determine a suitable subset $\mathcal{W}$ and how to compute $k(\mathcal{W})$. (Note that the latter problem is NP-hard, since it involves the solution of an albeit smaller instance of IOSANA on the subset $\mathcal{W}$.)

In order to generate a suitable inequality (6.41) as a cutting plane we make use of several different heuristics. In all of them, we first construct a candidate subset $\mathcal{W}$ and then compute the number of vehicles that is necessary to serve the trips in $\mathcal{W}$. Let $v^*, w^*, x^* \in [0,1]$ be the optimal solution of the LP relaxation of model (4.13). The heuristic strategies base on the observation that in the LP relaxation "fractional subcycles" frequently occur. By this we mean a subset of trips $t_1, \ldots, t_k$ where $x^*_{t_1 t_2}, x^*_{t_2 t_3}, \ldots, x^*_{t_k t_1}$ are close 1, i.e., to their upper bound.

Heuristic 1. As a seed we identify a deadhead trip $(t_1, t_2) \in \mathcal{A}$ with $(t_1, t_2) = \operatorname{argmax}\{x^*_{ij} : (i,j) \in \mathcal{A}\}$. We then start from t_2 and identify a trip t_3 with $(t_2, t_3) = \operatorname{argmax}\{x^*_{t_2 i} : (t_2, i) \in \mathcal{A}\}$. In this way we continue building up a sequence of trips $t_1, t_2, t_3, \ldots$ until either a circle occurs, that is, for some i, j with $i < j$ we have $t_i = t_j$ within the sequence, or the length of the sequence reaches an upper limit (say 10 trips, for instance). In the first case, subset $\mathcal{W}$ is the circle in the sequence. In the latter case, subset $\mathcal{W}$ is the entire sequence.

Heuristic 2. This heuristic is a slight modification of the previous one. As a seed we now take a deadhead trip $(t_1, t_2) \in \mathcal{A}$ with $(t_1, t_2) = \operatorname{argmin}\{f(\{i,j\}) : (i,j) \in \mathcal{A}\}$. We start with $\mathcal{W} := \{t_1, t_2\}$ and identify a trip t_3 with $t_3 = \operatorname{argmin}\{f(\mathcal{W} \cup \{i\}) : i \in \mathcal{V}\}$. If $f(\mathcal{W} \cup \{t_3\}) < f(\mathcal{W}) + \varepsilon$ then we let $\mathcal{W} := \mathcal{W} \cup \{t_3\}$. These steps are now repeated until no more trip is found or $|\mathcal{W}|$ has reached a given limit (10 trips, for instance). Here ε is a parameter that takes control of the generated set $\mathcal{W}$. A typical value we used is $\varepsilon := 0.5$.

Heuristic 3. We solve an auxiliary integer program to find a candidate set $\mathcal{W}$. For this we introduce an artificial depot node 0 and let $\mathcal{V}_0 := \mathcal{V} \cup \{0\}, \mathcal{A}_0 := \mathcal{A} \cup (\{0\} \times \mathcal{V}) \cup (\mathcal{V} \times \{0\})$. The decision variables $w \in \{0,1\}^{\mathcal{V}_0}$ indicate whether trip t is in $\mathcal{W}$ (for $w_t = 1$) or not (for $w_t = 0$). The decision variables $y \in \{0,1\}^{\mathcal{A}_0}$ indicate the paths flowing into $\mathcal{W}$. We have $y_{t_1 t_2} = 1$ if and only if $t_1 \notin \mathcal{W}$ and $t_2 \in \mathcal{W}$. The y and z variables are coupled through

$$y_{t_1 t_2} \leq 1 - z_{t_1}, \tag{6.42}$$

$$y_{t_1 t_2} \leq z_{t_2}, \tag{6.43}$$

$$z_{t_2} - z_{t_1} \leq y_{t_1 t_2} \tag{6.44}$$

for all $(t_1, t_2) \in \mathcal{A}_0$. The depot node 0 is only an artificial node. Hence it can never belong to $\mathcal{W}$:

$$z_0 = 0. \tag{6.45}$$

We restrict the maximum size of $\mathcal{W}$ by an upper bound constraint

$$\sum_{t \in \mathcal{V}} z_t \leq M, \tag{6.46}$$

(with $M := 10$, for instance). On the other hand, there is a minimum number of elements that should be in $\mathcal{W}$. Thus we have a lower bound constraint

$$\sum_{t \in \mathcal{V}} z_t \geq m. \tag{6.47}$$

The objective is to find a subset $\mathcal{W}$ with minimum inflow. The (heuristic) hope is that this would give a promising candidate for a k-path cut. As objective function we thus use

$$\sum_{t \in \mathcal{V}} v_t^* y_{0t} + \sum_{(t_1,t_2) \in \mathcal{A}} x_{t_1 t_2}^* y_{t_1 t_2} + \sum_{t \in \mathcal{V}} w_t^* y_{t0}. \tag{6.48}$$

Summing up, we solve the following integer program:

$$
\begin{aligned}
F_{m,M} = \quad &\min \quad &&(6.48)\\
&\text{subject to} \quad &&(6.42), \dots, (6.47)\\
& &&y \in \{0,1\}^{\mathcal{A}_0}, z \in \{0,1\}^{\mathcal{V}_0}.
\end{aligned} \tag{6.49}
$$

We solve (6.49) using a standard branch-and-cut MIP-solver to optimality. Denote y^*, z^* the optimal solution and let $\mathcal{W} := \{t \in \mathcal{V} : z_t^* = 1\}$. Then by definition we have $F_{m,M} = f(\mathcal{W})$.

For the second problem, the computation of $k(\mathcal{W})$, we solve an instance of IOSANA restricted to the set of trips $\mathcal{W}$ using the model (4.13) and a standard branch-and-cut MIP solver. Since we restrict ourselves to small sets $\mathcal{W}$ with typically no more than 10 trips, the solution can be obtained rather quickly (typically in less than one second). If we have $k(\mathcal{W}) > f(\mathcal{W})$, a violated k-path cut is found and added to strengthen the LP relaxation.

Combinatorial Benders' Cuts

Combinatorial Benders' cuts were introduced by Codato and Fischetti [20]. The central idea is to decompose the model into a master and a subproblem. The optimal solution of the master problem is then handed to the subproblem, where either feasibility is determined or a cutting plane for the master is returned. The decomposition is claimed to be especially useful for models that suffer from "big-M" constraints. Since we have such constraints in IOSANA, namely (4.8), this model fits to the scope of combinatorial Benders' cuts.

In the case of (4.13) the integer program decomposes into the *master*

$$
\begin{aligned}
&\min \quad &&((4.11),(4.12))\\
&\text{subject to} \quad &&(4.6),(4.7)\\
& &&v, w \in \{0,1\}^{|\mathcal{V}|}, x \in \{0,1\}^{|\mathcal{A}|},
\end{aligned} \tag{6.50}
$$

and an integer linear system parametrized by x

$$
\begin{array}{rl}
\text{subject to} & (4.8), (4.9), (4.10) \\
& (4.1), (4.2) \\
& \tau \in \mathbb{Z}^{|S|}, \alpha \in \mathbb{Z}^{|\mathcal{V}|},
\end{array} \tag{6.51}
$$

called the *slave*. We then solve the master problem to optimality and obtain a binary solution vector v^*, w^*, x^*. Afterwards we solve the slave with parameter x^* in inequality (4.8). If a feasible solution α^*, τ^* is found, $(v^*, w^*, x^*, \alpha^*, \tau^*)$ is an optimal solution for (4.13). If instead the slave is infeasible then x^* is already infeasible for problem (4.13). In this case we search an IIS for the slave system (6.51). Practically this is carried out by starting time propagation and algorithm findIIS from Section 2.2.6. We assume that the instance IOSANA was certified to have a feasible solution. Then the returned IIS of (6.51) contains at least one inequality of (4.8). Let $I \subset \mathcal{A}$ denote the index set corresponding to these deadhead trips. In order to break the infeasibility at least one binary variable x_{t_1, t_2} from I has to be 0. This condition is translated into the inequality

$$
\sum_{(t_1, t_2) \in I} x_{t_1 t_2} \leq |I| - 1, \tag{6.52}
$$

called *combinatorial Benders' cut*. For each given infeasible x^* one or more combinatorial Benders' cuts are generated and added to the master problem (6.50). Iterating this procedure finally produces an integer feasible and optimal solution of (4.13) after finitely many steps.

6.2 Modeling Alternatives

The IOSANA model presented in Chapter 4 is not the only way to formulate the problem as a (mixed) integer program. In this section we present and discuss two interesting alternatives.

6.2.1 Big-M-Free Reformulations

In [3] and [4], Ascheuer, Fischetti, and Grötschel (who give the credit to Maffioli and Sciomachen [53] resp. van Eijl [73]) describe a model that avoids the need of "big-M" terms (such as (4.8) in our IOSANA model). Their reformulation was introduced to solve instances of a single-vehicle traveling salesman problem with time windows. However, it can easily be extended to the case of multiple vehicles. We will now demonstrate how to apply their approach to a reformulation of our model (4.13).

We introduce additional integer variables $\xi_{t_1 t_2} \in \mathbb{Z}_+$ for all $(t_1, t_2) \in \mathcal{A}$ and $\zeta_t \in \mathbb{Z}_+$ for all $t \in \mathcal{V}$. These new variables represent the starting time of a trip in the following way:

If trip t_1 is connected with trip t_2 (that is, $x_{t_1 t_2} = 1$) then $\xi_{t_1 t_2}$ denotes the time when trip t_1 is starting. Otherwise if $x_{t_1 t_2} = 0$ then $\xi_{t_1 t_2} = 0$. If trip t is the last trip in some block (that is, $w_t = 1$) then ζ_t denotes the starting time of trip t. Otherwise if $w_t = 0$ then $\zeta_t = 0$. Mathematically this is modelled by the following inequalities.

The lower and upper bounds on the starting times for the ζ, ξ variables are coupled to the decision variables x and w. For all $(t_1, t_2) \in \mathcal{A}$ we have

$$\underline{\alpha}_{t_1} \cdot x_{t_1 t_2} \leq \xi_{t_1 t_2} \leq \overline{\alpha}_{t_1} \cdot x_{t_1 t_2} \tag{6.53}$$

and for all $t \in \mathcal{V}$ we have

$$\underline{\alpha}_t \cdot w_t \leq \zeta_t \leq \overline{\alpha}_t \cdot w_t. \tag{6.54}$$

In terms of the α variables the starting time of trip $t_1 \in \mathcal{V}$ is given by

$$\alpha_{t_1} = \zeta_{t_1} + \sum_{(t_1, t_2) \in \mathcal{A}} \xi_{t_1 t_2}. \tag{6.55}$$

Most important, inequalities (4.8) can be replaced by the following formulation without the use of "big-M"-terms for all $t_2 \in \mathcal{V}$

$$\sum_{t_1:(t_1,t_2)\in\mathcal{A}} \left(\xi_{t_1 t_2} + (\delta_{t_1}^{\mathrm{trip}} + \delta_{t_1 t_2}^{\mathrm{shift}} + \underline{\omega}_{t_1 t_2}^{\mathrm{idle}}) \cdot x_{t_1 t_2} \right) \leq \alpha_{t_2}, \tag{6.56}$$

and

$$\sum_{t_1:(t_1,t_2)\in\mathcal{A}} \left(\xi_{t_1 t_2} + (\delta_{t_1}^{\mathrm{trip}} + \delta_{t_1 t_2}^{\mathrm{shift}} + \overline{\omega}_{t_1 t_2}^{\mathrm{idle}}) \cdot x_{t_1 t_2} \right) \geq \alpha_{t_2}. \tag{6.57}$$

That is, if trip t_1 is connected with trip t_2 by a deadhead trip, then the starting time α_{t_2} of trip t_2 is bounded by the starting time $\xi_{t_1 t_2}$ of t_1 plus the driving time for the trip t_1 and the deadhead trip (t_1, t_2) and the minimum and maximum idle time, respectively. If t_1 is not connected with t_2, then the left-hand side of both inequalities (6.56) and (6.56) is equal to 0.

Using this big-M-free formulation, the bicriteria model (4.13) now looks as follows:

$$\begin{aligned}
\min \quad & ((4.11), (4.12)) \\
\text{subject to} \quad & (4.6), (4.7), (4.9), (4.10), (6.53), \ldots, (6.57) \\
& (4.1), (4.2) \\
& v, w \in \{0,1\}^{|\mathcal{V}|}, x \in \{0,1\}^{|\mathcal{A}|} \\
& \tau \in \mathbb{Z}^{|\mathcal{S}|}, \alpha, \zeta \in \mathbb{Z}^{|\mathcal{V}|}, \xi \in \mathbb{Z}^{|\mathcal{A}|}.
\end{aligned} \tag{6.58}$$

6.2.2   Set-Partitioning Based Relaxations

Balinski and Quandt [6] (see also the survey of Bramel and Simchi-Levi [13]) suggested a formulation for the classical capacitated vehicle routing problem that is based on a set partitioning problem. We adapt their approach to obtain strong lower bounds for instances of IOSANA.

We start with an enumeration of all feasible blocks. The feasibility of each block is checked by calling the starting time propagation. For the large real-world instances of IOSANA (see Chapter 7) the number of blocks ranges from 5 up to several million. Hence this enumerative phase is very time consuming, since the feasibility check involves each time the solution of an NP-hard problem with a pseudo-polynomial algorithm, the starting time propagation.

Denote by $\mathcal{R}$ the set of all feasible blocks. Its elements $r \in \mathcal{R}$ are sequences of the form $r = (t_{r_1}, t_{r_2}, \ldots, t_{r_k})$. That means, a bus drives from the depot to serve trip t_{r_1}, then drives a deadhead trip to the start of trip t_{r_2}, serves t_{r_2}, and so on, until finally the bus serves trip t_{r_k}, and drives back to the depot. When feasibility of block r is checked we fix the decision variables of the corresponding deadhead trips to their upper bounds 1, i.e., $x_{t_{r_1} t_{r_2}} = x_{t_{r_2} t_{r_3}} = \ldots = x_{t_{r_{k-1}} t_{r_k}} = 1$ and call the starting time propagation procedure thereafter.

allBlocks($\mathcal{I}$)

Input:	IOSANA instance $\mathcal{I}$
(1)	Let $i := 1, k_1 := 1, r := (), \mathcal{R} := \emptyset$
(2)	**Repeat**
(3)	Append trip t_{k_i} at the end of r
(4)	Call starting time propagation with r as fixed deadhead trips
(5)	**If** r is feasible **Then**
(6)	Let $i := i + 1$
(7)	Let $k_i := 1$
(8)	**Else**
(9)	Remove last trip in r
(10)	**Repeat**
(11)	Let $k_i := k_i + 1$
(12)	**Until** t_{k_i} is not in r and $k_i \leq |\mathcal{V}|$
(13)	**If** $k_i > |\mathcal{V}|$ **Then**
(14)	Let $\mathcal{R} := \mathcal{R} \cup \{r\}$
(15)	Remove last trip in r
(16)	Let $i := i - 1$
(17)	**End If**
(18)	**End If**
(19)	**Until** $i < 0$
Output:	set $\mathcal{R}$ of all feasible blocks for instance $\mathcal{I}$

We introduce some more notations. Let c_r be the cost of block $r \in \mathcal{R}$ (i.e., some suitable combination of the cost for the bus (4.11) and the length of the deadhead trips (4.12) in r). Define a matrix $A \in \{0,1\}^{\mathcal{V} \times \mathcal{R}}$ with $a_{tr} = 1$ if trip $t \in \mathcal{V}$ is included in block $r \in \mathcal{R}$ and $a_{tr} = 0$ otherwise. We introduce binary variables $y_r \in \{0,1\}$ for all blocks $r \in \mathcal{R}$ with $y_r = 1$ if block r is selected and $y_r = 0$ otherwise. Then in the set partitioning formulation of IOSANA the objective is to minimize the total costs of all selected blocks, where each trip must be included in some block. It is

$$z^{\text{sp}} = \min\{c^T y : Ay = \mathbf{1}, y \in \{0,1\}^{\mathcal{R}}\}, \tag{6.59}$$

where $\mathbf{1} := (1, 1, \ldots, 1)$. The computational problem is how to solve (6.59). The first idea is to consider the LP-relaxation

$$z^{\text{splp}} = \min\{c^T y : Ay = \mathbf{1}, y_r \geq 0 \ \forall \ r \in \mathcal{R}\}, \tag{6.60}$$

which already gives a lower bound $z^{\text{splp}} \leq z^{\text{sp}}$. However, solving (6.60) can also be quite difficult if the number of columns in A (i.e., the number of elements in $\mathcal{R}$) is too large (as it is in particular in the real-world instances of IOSANA).

For this reason we start with a small selection $\mathcal{R}' \subset \mathcal{R}$ of columns and solve the linear programming relaxation of (6.59) with respect to this reduced set:

$$z^{\text{splpr}} = \min\{c^T y : \sum_{r \in \mathcal{R}'} a_{tr} y_r = 1 \ \forall \ t \in \mathcal{V}, y_r \geq 0 \ \forall \ r \in \mathcal{R}'\}. \tag{6.61}$$

Denote y^* the optimal solution to problem (6.61) and let $\pi^* = (\pi_1^*, \ldots, \pi_{|\mathcal{V}|}^*)$ be the dual variables corresponding to the rows (constraints) in $Ax = 1$. To determine whether y^* is optimal for (6.60) we equivalently check whether π^* is optimal for the dual of (6.60). This linear program looks as follows:

$$\max\{\mathbf{1}^T \pi : \pi A \leq c\}. \tag{6.62}$$

Thus if π^* satisfies every constraint $\pi^* A \leq c$ then it is optimal for (6.62) and therefore y^* is optimal for (6.60). Otherwise if π^* is not feasible for (6.62) we can identify a violated constraint (a block) $r \in \mathcal{R} \backslash \mathcal{R}'$ (by sequentially checking all elements in this set) with $\pi^* a_r > c_r$. Block r is then added to $\mathcal{R}'$ and problem (6.61) is solved again. This procedure is repeated until no more violating constraint is found.

In case of vehicle routing problems known in literature so far, such as the capacitated VRP or the VRP with time windows, the feasibility of each solution can be determined by sequentially checking the tours for the individual vehicles: If each tour fulfills the required time windows and capacity constraints then the entire solution is declared feasible. In case of the VRPCTW (and thus IOSANA) this is no longer true. Due to the coupled time windows the feasibility of each single block does not imply the feasibility of the entire solution. This means that in general even feasible integer solutions of (6.59) cannot be interpreted as feasible solutions for (4.13).

In the sequel we restrict our attention to the problem how to obtain good feasible solutions for (4.13) from solutions of

$$z^{\text{spr}} = \min\{c^T y : \sum_{r \in \mathcal{R}'} a_{tr} y_r = 1 \ \forall \ t \in \mathcal{V}, y \in \{0,1\}^{\mathcal{R}'}\} \tag{6.63}$$

instead of (6.59), where $\mathcal{R}'$ is a subset of columns of $\mathcal{R}$ such that no column in $\mathcal{R} \backslash \mathcal{R}'$ has negative reduced costs. In particular we cannot guarantee global optimality. We now present some cutting planes which are able to cut off infeasible combinations of columns (i.e., blocks) in (6.63).

Reactivating the Model Inequalities

We use constraints from the original model (4.13) for (6.63) to enforce feasibility in the selection of blocks. For this, we add the variables α_t for $t \in \mathcal{V}$ and τ_s for $s \in \mathcal{S}$ and their bounds (4.2) and (4.1) to the set partitioning model. Constraints (4.9) and (4.10) can also be incorporated directly. Constraints (4.8) have to be altered somehow since the x variables are not part of the set partitioning formulation.

For every column $r = (t_{r_1}, \ldots, t_{r_k}) \in \mathcal{R}'$ we introduce the inequalities

$$
\begin{aligned}
\alpha_{t_{r_i}} + \delta_{t_{r_i}}^{\text{trip}} + \delta_{t_{r_i} t_{r_{i+1}}}^{\text{shift}} + \underline{\omega}_{t_{r_i} t_{r_{i+1}}}^{\text{idle}} - M'_{ri} \cdot (1 - y_r) & \leq \alpha_{t_{r_{i+1}}}, \\
\alpha_{t_{r_i}} + \delta_{t_{r_i}}^{\text{trip}} + \delta_{t_{r_i} t_{r_{i+1}}}^{\text{shift}} + \overline{\omega}_{t_{r_i} t_{r_{i+1}}}^{\text{idle}} + M''_{ri} \cdot (1 - y_r) & \geq \alpha_{t_{r_{i+1}}}
\end{aligned}
\tag{6.64}
$$

for every $i \in \{1, \ldots, k-1\}$.

Again, M'_{ri} and M''_{ri} are sufficiently big values, depending on r and i. According to (6.9) and (6.10) we can set $M'_{ri} := M'_{t_{r_i} t_{r_{i+1}}}$ and $M''_{ri} := M''_{t_{r_i} t_{r_{i+1}}}$ for all $i \in \{1, \ldots, k-1\}$ to strengthen (6.64) as far as possible.

Combinatorial Benders' Cuts

In Section 6.1.2 we presented the general idea of combinatorial Benders' cuts and applied it to the bicriteria model (4.13). The same can be done for the set partitioning formulation, as we will see now.

Let $y^* \in \{0, 1\}^{\mathcal{R}'}$ be an integer optimal solution of the set partitioning relaxation (6.63). By the starting time propagation algorithm it can be easily checked whether the subset $\mathcal{R}^* := \{r \in \mathcal{R}' : y_r^* = 1\}$ corresponds to a feasible schedule or not. If it does, we are done: We found the best schedule within the restricted subset of columns. If not, we identify an irreducible infeasible subset (IIS) of $\mathcal{R}^*$, i.e., a subset $\mathcal{R}^{\text{IIS}} \subseteq \mathcal{R}^*$ with the property that the schedules of $\mathcal{R}^{\text{IIS}}$ are infeasible, but for all $r \in \mathcal{R}^{\text{IIS}}$, the schedules $\mathcal{R}^{\text{IIS}} \setminus \{r\}$ are feasible. Such "column-IIS" can be identified in pseudo-polynomial time using the starting time propagation algorithm. If $\mathcal{R}^{\text{IIS}}$ is a column-IIS then it gives rise to the following *combinatorial Benders' cut*

$$
\sum_{r \in \mathcal{R}^{\text{IIS}}} y_r \leq |\mathcal{R}^{\text{IIS}}| - 1,
\tag{6.65}
$$

which says that at least one column from $\mathcal{R}^{\text{IIS}}$ must be inactive in a feasible solution.

We remove the columns of $\mathcal{R}^{\text{IIS}}$ from $\mathcal{R}^*$ and repeat the search for another IIS, until no more IIS is found. All corresponding inequalities are added to the LP within a branch-and-cut framework.

Aggregated Combinatorial Benders' Cuts

In principle we could add combinatorial Benders' cuts for the columns in $\mathcal{R}'$ in advance. Practically this is not possible, because of the high number of inequalities. However, if we restrict ourselves to IIS with two columns and aggregate all inequalities belonging to the same column, we only get one additional inequality per column.

Since it is still very time consuming to check the infeasibility of each pair of columns, we use a sufficient condition for infeasibility instead. For each column $r \in \mathcal{R}'$ we fix the corresponding deadhead trips. Then we run the starting time propagation to obtain new bounds $\underline{\alpha}^r \leq \alpha \leq \overline{\alpha}^r$ and $\underline{\tau}^r \leq \tau \leq \overline{\tau}^r$ for the trip and school starting time windows, respectively. These bounds clearly depend on the column r. Now if there are two distinct columns $r, q \in \mathcal{R}'$ and a school s with $\underline{\tau}_s^r > \overline{\tau}_s^q$ or a trip t with $\underline{\alpha}_t^r > \overline{\alpha}_t^q$ then column r, q are within an IIS, because the time windows are incompatible.

For a given column $r \in \mathcal{R}'$ we identify in this manner all columns with incompatible time windows. The subset of these columns is denoted as $\mathcal{R}_r^{\text{INC}} \subseteq \mathcal{R}'\backslash\{r\}$. Then the following so-called *aggregated combinatorial Benders' cut* can be added to (6.63)

$$\sum_{q \in \mathcal{R}_r^{\text{INC}}} y_q \leq |\mathcal{R}_r^{\text{INC}}|(1 - y_r), \qquad (6.66)$$

meaning that either y_r is in the solution or some columns from $\mathcal{R}_r$.

Set-Covering Type Inequalities

We now introduce a class of inequalities that are in general stronger than the aggregated combinatorial Benders' cuts.

Similar to the aggregated combinatorial Benders' cuts we run the starting time propagation for each column $r \in \mathcal{R}'$. Instead of excluding columns as we did in inequality (6.66) we now use the coupled time windows to influence the selection of columns.

Consider some column $r = (t_{r_1}, t_{r_2}, \ldots, t_{r_k}) \in \mathcal{R}'$ and assume that for some $i \in \{1, \ldots, k\}$ trip t_{r_i} transfers pupils to some school s, i.e., $(s, t_{r_i}) \in \mathcal{P}$. Then for every other trip $t \in \mathcal{V}\backslash\{t_{r_i}\}$ with $(s, t) \in \mathcal{P}$ we identify the subset of columns $\mathcal{R}(t) := \{q \in \mathcal{R}' : t \in q\}$. This set $\mathcal{R}(t)$ can be divided into two disjoint subsets $\mathcal{R}(t) = \mathcal{R}_r^{\text{COM}}(t) \cup \mathcal{R}_r^{\text{INC}}(t)$, where $\mathcal{R}_r^{\text{COM}}(t)$ consists of those columns with time windows compatible to r and $\mathcal{R}_r^{\text{INC}}(t)$ of those columns with incompatible time windows with respect to r. From this we get the *set-covering type inequality*

$$y_r \leq \sum_{q \in \mathcal{R}_r^{\text{COM}}(t)} y_q. \qquad (6.67)$$

We theoretically compare the strength of inequalities (6.66) with (6.67). From the corresponding set partitioning equality in (6.63) we know that exactly one column in $\mathcal{R}(t)$ has to be chosen. That is,

$$1 = \sum_{q \in \mathcal{R}(t)} y_q = \sum_{q \in \mathcal{R}_r^{\text{COM}}(t)} y_q + \sum_{q \in \mathcal{R}_r^{\text{INC}}(t)} y_q. \tag{6.68}$$

Combining (6.68) with (6.67) we obtain

$$\sum_{q \in \mathcal{R}_r^{\text{INC}}(t)} y_q \leq 1 - y_r. \tag{6.69}$$

Inequality (6.66) in this case reads

$$\sum_{q \in \mathcal{R}_r^{\text{INC}}(t)} y_q \leq |\mathcal{R}_r^{\text{INC}}(t)|(1 - y_r), \tag{6.70}$$

hence for $|\mathcal{R}_r^{\text{INC}}(t)| > 1$ inequality (6.67) is superior compared to the aggregated combinatorial Benders' cut (6.66).

Clique Inequalities

Our final class of cuts are *clique cuts*, where a clique is (heuristically) determined in some conflict graph that gives rise to a valid inequality. Clique cuts were first described by Fulkerson [33] and Padberg [60]. The conflict graph $G = (V, E)$ is constructed as follows. For every column $r \in \mathcal{R}'$ introduce a node in V (hence there is a one-to-one correspondence between the columns in $\mathcal{R}'$ and the nodes in V). We now proceed as in the aggregated combinatorial Benders' cuts: For each column $r \in \mathcal{R}'$ we fix the corresponding deadhead trips and run the starting time propagation. If two columns r, q have incompatible time windows (see above), then an edge between the nodes corresponding to r and q is introduced in G.

Then we seek a clique in G, i.e., a subset of nodes $V' \subseteq V$ such that each node $v \in V'$ is adjacent to all other nodes in V'. The separation of clique cuts is difficult (*NP*-hard), see Groetschel, Lovasz, and Schrijver [37], thus the construction of a clique cut is actually done by a greedy-type heuristic. This heuristic starts at some node v and sets $V' := \{v\}$. Then iteratively nodes are added to V' that are adjacent to all other nodes that are already in V', until no further node is found. This procedure is then repeated a limited number of times for different start nodes v.

Each clique V' gives rise to the following valid inequality:

$$\sum_{r \in V'} x_r \leq 1. \tag{6.71}$$

6.3 Combining Primal and Dual Solution Techniques

Although the pgreedy heuristic alone produces already reasonable results, it also can get stuck in local-optimal solutions, if the parameters within the greedy heuristic are not approximating the solution space well enough. In this case the heuristic is not able to find feasible solutions close to the globally optimal solution. In this section we describe how the set partitioning model can be combined with the pgreedy heuristic to obtain even better solutions.

From the heuristic's point of view the selection process of new columns from the column pool can be seen as an additional local search feature. It provides schedules which the heuristic is not able to produce itself. From the column generation's point of view the solutions of the heuristic can be interpreted as a set of additional columns so that in each step not only one column enters the matrix, but a whole set of columns, all belonging to a local-optimal solution.

Assume that all blocks are enumerated and stored in a column pool $\mathcal{R}$. We start with the pgreedy heuristic. When it terminates (note that it is sufficient to run stage one), a feasible solution in terms of bus schedules is produced. We then identify the columns in $\mathcal{R}$ that correspond to the blocks in this solution. These columns are stored in $\mathcal{R}'$. Then we solve the LP relaxation of the set partitioning problem (6.61) and seek for a column $r \in \mathcal{R} \backslash \mathcal{R}'$ with $\pi^* a_r > c_r$. If there is such a column we fix the deadhead trips corresponding to r and call the heuristic (stage one) with these fixings again. Then a new feasible solution is returned and the columns corresponding to this new solution are added to $\mathcal{R}'$ (in case they were not already in $\mathcal{R}'$).

These steps are now iterated until $\pi^* a_r \leq c_r$ for all $r \in \mathcal{R} \backslash \mathcal{R}'$. Then we end up with a subset of columns $\mathcal{R}'$ from the different solutions found by the pgreedy heuristic in the intermediate steps.

In a final step we (try to) solve (6.63) using the inequalities described above to obtain an even better solution from the combination of blocks from different feasible solutions.

Chapter 7

Input Data for the Model

An instance of the problem is described by a bundle of sets and parameters as described in Section 4.1. In this chapter we discuss how these are obtained for our set of test instances. Mainly, we have two sources: Randomly generated sets of various sizes, and real-world instances from counties where our project partner BPI-Consult actually worked in the past few years and gathered the data. We start in Section 7.1 with a manually constructed instance called "the tiny one". In Section 7.2 we present the five counties and discuss how the input data is transformed into problem data. For the generation of random data we present in Section 7.3 a method to obtain instances with the property that they somehow "look" realistic.

7.1 A Very Tiny Instance

In the sequel we present a very small instance. It was generated as the demonstration example for our patent [31].

In this instance there are six trips, $\mathcal{V} := \{t_1, t_2, t_3, t_4, t_5, t_6\}$, with the following parameters:

	t_1	t_2	t_3	t_4	t_5	t_6
δ_t^{trip}	30	30	30	40	30	60
$\hat{\alpha}_t$	7:20	7:30	7:40	7:30	7:10	7:00
$\underline{\alpha}_t$	6:30	6:30	6:30	6:30	6:30	6:30
$\overline{\alpha}_t$	8:30	8:30	8:30	8:30	8:30	8:30

$$(7.1)$$

Deadhead trips are possible between all pairs of trips, hence $\mathcal{A} := \{(t_i, t_j) \in \mathcal{V} \times \mathcal{V} : i \neq j\}$. We assume that there is no minimum and maximum idle time after a deadhead trip, i.e., $\underline{\omega}_{t_i t_j}^{\text{idle}} = 0$ and $\overline{\omega}_{t_i t_j}^{\text{idle}} = \infty$ for all $(t_i, t_j) \in \mathcal{A}$. We have the following driving times $\delta_{t_i t_j}^{\text{shift}}$

for the deadhead trips:

	t_1	t_2	t_3	t_4	t_5	t_6
t_1	–	10	5	20	25	20
t_2	60	–	10	5	20	15
t_3	40	50	–	15	10	15
t_4	30	55	35	–	10	10
t_5	35	65	40	50	–	10
t_6	40	70	55	70	10	–

$$(7.2)$$

There are three schools, $\mathcal{S} := \{s_1, s_2, s_3\}$, with the following parameters:

	s_1	s_2	s_3
$\hat{\tau}_s$	8:00	8:10	7:50
$\underline{\tau}_s$	7:30	7:30	7:30
$\overline{\tau}_s$	8:30	8:30	8:30

$$(7.3)$$

Let $\mathcal{P} := \{(s_1, t_2), (s_1, t_3), (s_2, t_3), (s_2, t_4), (s_3, t_5), (s_3, t_6)\}$, that is, the trips t_2, t_3 transport pupils to school s_1, trips t_3, t_4 to school s_2, and trips t_5, t_6 to school s_3, with parameters as follows::

	(s_1, t_2)	(s_1, t_3)	(s_2, t_3)	(s_2, t_4)	(s_3, t_5)	(s_3, t_6)
$\delta_{st}^{\text{school}}$	20	10	20	15	10	20
$\underline{\omega}_{st}^{\text{school}}$	5	5	5	5	5	5
$\overline{\omega}_{st}^{\text{school}}$	45	45	45	45	45	45

$$(7.4)$$

We have only one pair of feeder and collector trips, $\mathcal{C} := \{(t_1, t_2)\}$, i.e., trip t_1 is a feeder trip for collector trip t_2 with parameters $\delta_{t_1 t_2}^{\text{feeder}} := 20, \delta_{t_1 t_2}^{\text{collector}} := 10, \underline{\omega}_{t_1 t_2}^{\text{collector}} := 0, \overline{\omega}_{t_1 t_2}^{\text{collector}} := 10$. There are no free trips.

Because of the trip starting times it is not possible to connect any two trips within a schedule. In other words, for every trip we have to use a new bus, thus use 6 buses altogether. The "county" is shown in Figure 7.1. In the next chapter we demonstrate that a suitable change of the schools' and trips' starting times can reduce the number of buses from 6 to 2.

7.2   Real-World Data

Real-world data for the model was available from five different German counties. Below we present the counties, their geography and history in greater detail. In the beginning, the data was not in the format that was needed. In this section we describe how we extracted all necessary information, i.e., how real-world *raw data* is transformed into *input data* for the model (i.e., the sets and parameters described in Section 4.1).

Common to all five raw data sets is the availability of two sets of raw data. First, there are the time-tables of all trips that are actually served in the county. For each trip, the

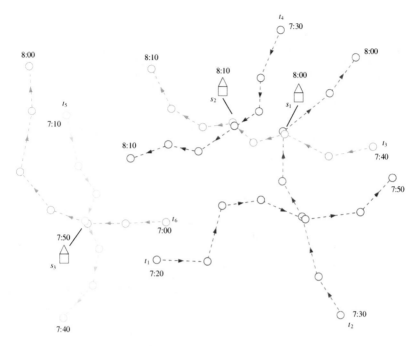

Figure 7.1: A tiny example

time-table is a list of bus stops and the times when the bus arrives and departs at each bus stop. Second, there are lists of schools with their actual starting times and the bus stops they are assigned to, i.e., where their pupils leave the bus.

From the time-tables, the following sets and parameters are directly determined: The set $\mathcal{V}$, the current trip starting time $\hat{\alpha}_t$, and the time for serving the entire trip δ_t^{trip}. From the list of schools, we obtain the set $\mathcal{S}$, and the current school starting time $\hat{\tau}_s$. All other sets and parameters are not contained in the input data, and thus have to be generated in a way we describe now.

Bounds on the school starting time $\underline{\tau}_s, \overline{\tau}_s$ are in general determined by legal restrictions (for instance, 7:30 – 8:30). However, for some reason the current school starting time is sometimes outside these legal bounds. In this exceptional case, the bounds on the school starting times are relaxed, so that the current starting time is inside the bounds. For instance, in County Demmin there is a school that currently starts at 7:10. Hence, the starting time window for this particular school is set to 7:10 – 8:30.

Bounds on the trip starting times $\underline{\alpha}_t, \overline{\alpha}_t$ are depending on the type of the particular trip. That is, all school trips, feeder and collector trips have rather weak bounds: $\underline{\alpha}_t := 0, \overline{\alpha}_t := 1439$, meaning that they have to start somewhen between midnight and 23:59. These weak bounds are later adjusted by the starting time propagation subroutine. Only the free trips get tighter bounds right from the start, because they are not changed by the starting time propagation. Here one typically allows the trip starting times to change by plus or minus 20 minutes relatively to the current starting times, i.e., $\underline{\alpha}_t := \hat{\alpha}_t - 10, \overline{\alpha}_t := \hat{\alpha}_t + 10$.

The set of deadhead trips $\mathcal{A}$ is the set of all pairs of trips (t_1, t_2) in $\mathcal{V} \times \mathcal{V}$ with $t_1 \neq t_2$, hence it has $n \cdot (n - 1)$ elements. The minimum idle times for buses between two consecutive trips within a block is in general set to 2 minutes, whereas there is no upper bound on this, i.e., $\underline{\omega}_{t_1 t_2}^{\text{idle}} := 2, \overline{\omega}_{t_1 t_2}^{\text{idle}} := \infty$. It is more difficult to determine a sensible value for the deadhead trip duration $\delta_{t_1 t_2}^{\text{shift}}$. We did not want to make use of external (as thus expensive) sources such as route planning tools. Thus, the only possible way to obtain travel times for deadhead trips is to use the data that is already available from the time-tables of the trips. The assumption we make hereby is that the buses use for deadhead trips the same roads and same times that they use for the passenger trips.

We construct a directed graph $G = (V, A)$ with node set V and arc set A. For every bus stop in the time-table trips introduce a vertex in V. We introduce an arc (v_1, v_2), if there exists a bus that drives from the bus stop corresponding to vertex v_1 to v_2. The weight of these arcs is the driving time of the bus between this two bus stops. If there is some other bus driving from v_1 to v_2 with a different driving time, then we assign the minimum of both as a weight for arc (v_1, v_2). The time $\delta_{t_1 t_2}^{\text{shift}}$ for the deadhead trip between two trips $(t_1, t_2) \in \mathcal{A}$ is then the shortest path in G from the vertex corresponding to the last bus stop of t_1 to the vertex corresponding to the first bus stop of t_2. For the computation of all shortest paths we use Dijkstra's shortest path algorithm. The so-computed value is an estimation of the real traveling time for two reasons: First, a bus can drive at least as fast on a deadhead trip compared to the passenger trips. Second, a bus might use a shorter way for the deadhead trip that is not part of any passenger trips. Thus, in reality the deadhead time is less or equal to our estimated deadhead time. For an optimization tool this is not a major drawback, since it guarantees that any mathematically feasible solution is also feasible in practice. In all real-world instances we examined so far there was no information about the location of the depots available. Hence we set $\mathcal{D} = \emptyset$ and $\delta_t^{\text{out}} = \delta_t^{\text{in}} = 0$ for all $t \in \mathcal{V}$.

Next, the set $\mathcal{P}$ of school-trip pairings is determined. Again, this data was not provided as input data. In fact it is not uncommon that neither the bus companies know which schools the pupils in some bus attend, nor the schools know which buses their pupils actually take. For every school $s \in \mathcal{S}$ it is reasonable to expect that a bus carries pupils for this school, if the bus stops at one of the school's bus stops within some time window before the start of the school. The lower bound of this time window reflects the minimum walking time from the bus stop to school (for instance, 5 to 10 minutes), whereas the upper bound is due to legal restrictions (for example, 30 minutes for pupils in primary schools and 60 minutes for pupils in high schools). Every trip $t \in \mathcal{V}$ with this property is considered as a school trip, and we put (s, t) into the set $\mathcal{P}$. Mostly, $\delta_{st}^{\text{school}}$ then is uniquely determined as the travel time for bus t from the departure at the first bus stop of its time-table to the arrival at one of the school's bus stops. A few buses arrive more than once at the same school. In this case, we take the last arrival.

The bounds on the waiting time for pupils at their school before the beginning of the lessons, $\underline{\omega}_{st}^{school}$ and $\overline{\omega}_{st}^{school}$, are in general corresponding to the minimal walking time from the bus stop to the school, and the maximal legal bound, respectively. Because of quality-of-service aspects, a feasible solution of the model is not applicable in practice, if there is a too large gap between the waiting time difference of the current and the planned waiting times. On the other hand, a reduction of waiting times is positive for the pupils, but if the waiting time of only some pupils is reduced, and those of others is not, the solution is again declined due to political reasons. This means that the values of $\underline{\omega}_{st}^{school}$ and $\overline{\omega}_{st}^{school}$ are adjusted in such way, that the current and the planned waiting times do not differ too much.

Finally, the set $\mathcal{C}$ of feeder and collector trip pairings is computed. In the set $\mathcal{P}$ we have all pairs of trips and schools. In the same fashion we determine if two trips t_1 and t_2 exchange customers. Since we concentrate the whole optimization on pupils, we only identify those transfers where pupils leave one bus and continue their journey to school with some other bus. For every trip t_2 and every school s with $(t_2, s) \in \mathcal{P}$, we identify those trips t_1 in $\mathcal{V}$ that have a common bus stop with t_2. We then check the following conditions:

- The school bus stop of s is in the time table of t_2 behind the common bus stop of t_1 and t_2.

- Trip t_1 itself does not stop at school s, that is, $(t_1, s) \notin \mathcal{P}$. (If it does, why would anyone leave the bus?)

- For every bus stop of t_1 before the change bus stop, there is no school bus that would take the pupils directly to school. (If there is such a bus, then every pupil would use it instead of changing the bus.)

- For every bus stop of t_1 before the change bus stop, there is no other pair of feeder and collector trips that would take the pupils to school in a shorter amount of total driving time. (If there is such a pair, then pupils would prefer it.)

- The arrival and departure of t_1 and t_2 at this bus stop must lie in a small time window, such that changing the bus from t_1 to t_2 is possible (for instance, a time window from 0 to 5 minutes).

If t_1 and t_2 fulfill all of these properties, then it is reasonable to consider t_1 as a feeder and t_2 as a collector trip, and add (t_1, t_2) to the set $\mathcal{C}$. The starting times of t_1 and t_2 must then be synchronized, such that the change is still possible after the optimization. If t_1 and t_2 only have one bus stop in common, then the driving times for the buses serving feeder and collector trips from their respective first bus stops to this common bus stop,

Table 7.1: Sizes of the sets and the IP for the five real-world instances.

County	$|\mathcal{V}|$	$|\mathcal{A}|$	$|\mathcal{S}|$	$|\mathcal{P}|$	$|\mathcal{C}|$	variables	constraints
Demmin	247	60,762	43	195	165	61,589	62,019
Steinfurt	490	239,610	102	574	406	241,284	242,652
Soest	191	36,290	82	294	182	37,027	37,706
Wernigerode	134	17,822	37	201	204	18,298	18,937
Guetersloh	404	162,812	84	579	708	166,108	166,194

$\delta_{t_1 t_2}^{\text{feeder}}$ and $\delta_{t_1 t_2}^{\text{collector}}$, are uniquely determined. If there is more than one possibility, then we choose the latest possible.

Looking at feasible solutions of the model from a political point of view, the bounds on the waiting time for pupils between leaving the feeder trip and being picked up by the collector trip, $\underline{\omega}_{st}^{\text{change}}$ and $\overline{\omega}_{st}^{\text{change}}$, are not as critical as the bounds for the waiting time at the schools. Thus, one can simply set them to some reasonable values, say 0 to 5 minutes, without a comparison between current and planned waiting times at this stage.

The number of pupils using a certain trip $\varphi_{st}^{\text{school}}$ was not available for our five test instances. The same holds for the number of pupils transferring between two trips $\varphi_{t_1 t_2}^{\text{change}}$. Thus we set them constantly to 1, which means that in the multicriteria model of the optimization problem, each bus has the same weight, regardless how many pupils are actually affected by a change of its starting time.

Also the parameters $\hat{v}, \hat{w}, \hat{x}$ representing the current bus schedules in the counties were not avaible. Hence we neglect goal (4.27) from the further discussions.

We have five real-world instances at hand. Table 7.1 provides an overview on the sizes of their sets. The number of trips in set $\mathcal{V}$ ranges from 134 in Wernigerode, the smallest instance, to 490 trips in Steinfurt, the largest one. The number of deadhead trips in set $\mathcal{A}$ is just resulting as $|\mathcal{V}|^2 - |\mathcal{V}|$, since each trip can be connected with each other trip except the trip itself. The number of schools in $\mathcal{S}$ varies from 37 in the smallest county to 102 in the largest. The trips with pupils and the trips with transferring customers were generated automatically. They range from 195 to 577 trips with pupils in the set $\mathcal{P}$ and from 182 to 701 transfers in the set $\mathcal{C}$. In the last two columns of Table 7.1 the size of the corresponding integer programs (4.13) is given, where "size" means the number of variables and constraints.

We now present the counties in detail. The historical and geographical information and the coats of arms were taken from Wikipedia, the free encyclopedia [75].

7.2.1   Demmin

Demmin is a district in Mecklenburg-Western Pomerania, Germany. The following map
shows the villages and cities of county Demmin (small and big circles, respectively) and
the streets inbetween them. Each of the approximately 300 circles represents between
one and up to fifty (in Neubrandenburg) bus stops.

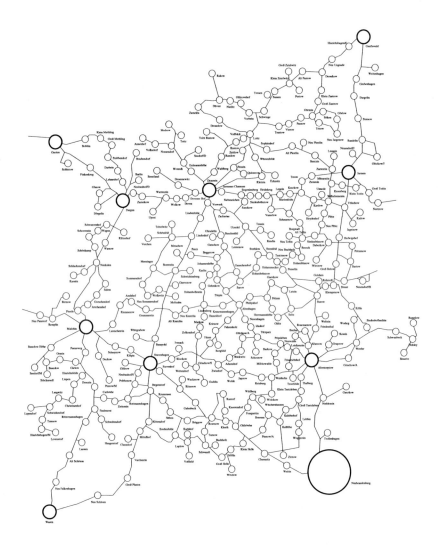

The county has 92,935 inhabitants (by the end of 2001) and an area of 1,921 km^2. It is bounded by (from the south and clockwise) the districts of Müritz, Güstrow, Nordvorpommern, Ostvorpommern, and Mecklenburg-Strelitz, and by the city of Neubrandenburg. The district was established in 1994 by merging the former districts of Demmin, Altentreptow and Malchin. The district of Demmin unites parts of the historical regions of Mecklenburg and (Western) Pomerania. In the west the Peene river enters the district, runs through a large lake called the Kummerower See (33 km^2), and leaves the district to the northeast.

The bull in the coat of arms is the heraldic animal of Mecklenburg , the silver griffin together with the chess pattern have been the arms of Pomerania, and the silver castle to the right is from the arms of the city of Demmin.

7.2.2   Steinfurt

Steinfurt is a Kreis (district) in the northern part of North Rhine-Westphalia, Germany. The county has 438,765 inhabitants (in 2002) and an area of 1,792 km^2. Neighboring districts are Bentheim, Emsland, district-free Osnabrück and the Osnabrück district, Warendorf, district-free Münster, Coesfeld, Borken. The district is situated at the Lower Saxonian border, north of Münster. The Ems river runs through the district from south to north. Highest elevation is the Westerbecker Berg with 234m, lowest point is the Bentlage castle at 32m. In late medieval times Steinfurt became an independent earldom. Originally it was a part of the earldom of Bentheim, before it became independent in 1454. 1804 Steinfurt was reunited with Bentheim, before it became a part of the Prussian province of Westphalia in 1815. The new government in 1816 created the districts Steinfurt and Tecklenburg. In 1975 the old district Steinfurt was merged with the district Tecklenburg, and together with Greven and Saerbeck from the former district Münster the current district was formed.

The coat of arms combines elements from the coat of arms of the former districts Steinfurt, Tecklenburg and Münster. The swan in the middle comes from the earldom Steinfurt, the center of the district. The red bar around the swan symbolizes the clerical state Münster, which lay around the dukedom Steinfurt. The red water lily leaves were symbol of the dukes of Tecklenburg.

7.2.3 Soest

Soest is a Kreis (district) in the northern part of North Rhine-Westphalia, Germany. The county has 307,809 inhabitants (in 2002) and an area of 1,327 km². Geographically it covers the northern part of Sauerland hills and the lower land north of it, the Hellweg. The rather flat land and very fertile loess soil makes it an old agricultural area. The main rivers through the district are the Ruhr, Lippe and the Möhne, which also forms the biggest artificial lake of North Rhine-Westphalia. The highest elevation is in near Warstein with 580m, the lowest with 65m is the Lippe valley. In medieval times Soest was the biggest city of Westphalia, however after it liberated itself from the bishops of Cologne in the Soester Fehde in 1449 it slowly lost importance, only to gain it again when in 1816 the new Prussian government created the district. In 1974 the district was merged with the neighboring district Lippstadt.

The coat of arms show two symbols from the bishops Cologne, who owned the Soest until 1449. In left half is the key of Saint Peter, the patron of Cologne, in the right the black cross of Cologne. After the district was merged with Lippstadt the rose as symbol of the Lippe area was added on top of the cross. This new version of the coat of arms was granted in 1976.

7.2.4   Wernigerode

Wernigerode is a district (Kreis) in the west of Saxony-Anhalt, Germany. The county has 94,556 inhabitants (in 2002) and an area of 798 km^2. Neighboring districts are (from north clockwise) Halberstadt, Quedlinburg, the district Nordhausen in Thuringia and the district Goslar in Lower Saxony. The district is located at the northern end of the Harz mountains. The highest mountain of the Harz, The Brocken, is located in the southeast of the district.

The coat of arms are based upon those used by the principality and the diocese of Halberstadt. On top of the red and white areas are two fish, which are taken from the coat of arms of the counts of Wernigerode.

7.2.5   Gütersloh

Gütersloh is a Kreis (district) in the north-east of North Rhine-Westphalia, Germany. The county has 345,370 inhabitants (in 2001) and an area of 967 km^2. Neighboring districts are Osnabrück, Herford, district-free Bielefeld, Lippe, Paderborn, Soest and Warendorf. The east of the district is covered by the Teutoburg Forest, which also contains the highest elevation of the district, the Hengeberg (316m). In the west there is the source of the Ems river. The Ems valley also contains the lowest point of the district, near Harsewinkel with 56m. The county was created in 1973 in the reorganization of the districts in North Rhine-Westphalia, when the previous districts of Halle and Rheda-Wiedenbrück were merged. Also the districts Bielefeld, Paderborn, Beckum, and Warendorf had to cease parts of their area to the newly formed district. The precursor districts were created in 1816 when the new Prussian province of Westphalia was established.

The coat of arms was adopted with the creation of the district in 1973 by combining the symbols from the arms of the previous districts Halle and Rheda-Wiedenbrück. The three elements derive from the medieval counties in the area now covered by the district. The golden eagle in the bottom of the coat comes from the duchy of Rietberg. The wheel in the middle derives from the sign of the city Osnabrück, which owned the Amt Reckenberg. The chevrons in the top part are the symbol of the duchy of Ravensberg.

7.3 Random Data

As one can imagine, transforming input data into problem data is a complicated and time consuming process, mostly because it involves several steps of digitalization and data correction that can only be carried out manually. In addition, the size of the instance is not flexible. It is either a big county or a small. For algorithm development and testing purposes it is desirable to have instances with designable properties at hand. Given a number of schools, trips, bus stops, etc., create an instance with exactly these characteristics. In addition, the produced instance should not look too artificial. In the sequel we describe how this can be achieved.

In contrast to the direct and somehow uncontrolled generation of random numbers as input figures, the main idea behind our generator of instances is to produce an entire county randomly, with smaller and bigger cities, streets between them, high schools in big cities, primary schools in the villages, and bus trips between cities and villages. Once this is done, the necessary sets and parameters can be derived the same way as for real-world instances, as described in Section 7.2.

There are two numbers that control the size of the generated instance: The number of trips T and the number of schools S. As in reality, a trip shall have from 10 to 30 bus stops and a duration varying from 20 to 60 minutes. For a given number of trips T, we generate about $10 \cdot T$ bus stops. This ensures, that in average every bus stop is visited by more than one bus.

A bus stop is a point in a two dimensional square $Q := [0, 1]^2$, the county. The bus stops are not uniformly distributed in the county. They are more dense in the cities than in the countryside. So we first randomly choose some city centers (for example,

Table 7.2: Sizes of the randomly generated input data sets and MIP models

	$|\mathcal{V}|$	$|\mathcal{A}|$	$|\mathcal{S}|$	$|\mathcal{P}|$	$|\mathcal{C}|$	variables	constraints
rnd_1	25	600	10	36	19	685	770
rnd_2	25	600	10	35	15	690	758
rnd_3	25	600	10	37	17	685	760

we took one center for every 10 trips). Now the bus stops are located in the following way. Select randomly (uniform distribution) a center c. Generate a random direction vector d uniformly distributed on the unit circle. Generate the bus stop coordinates by sampling uniformly over the set of points $\{x \in Q : x = c + \lambda \cdot d, \lambda > 0\}$. This assures that a bus stop is more likely situated near the center than in the periphery. As travel time we use the Euclidean distance between the points, scaled in such a way that the average travel time between two inner city bus stops is around 1 to 2 minutes.

Then the S schools are equally distributed among all bus stops. Thus we naturally obtain more schools in the cities than in the countryside, which is also a realistic assumption. The starting time for each school is set to 8:00, with a possible starting time between 7:30 and 8:30. In reality, the school starting time distribution was never that sharply focused, but here we want to give our algorithms something to play with.

The most interesting part is the generation of the trips. Every school has to be visited by at least one trip. For every trip, select a school. The school should be placed within the last third of the trip's time table: Beginning with the school's bus stop, the trip is thus growing in two directions. The selection of the next bus stop is biased by the distance and how often this bus stop was selected before, $\frac{b_i}{dist(i,j)}$. In the beginning, all bus stops i are assigned a factor $b_i = 1$. Every time some bus stop i is chosen for a trip, this factor is multiplied by some constant θ. After 20 bus stops are selected this way, a local 2-opt is used to reduce the travel time of the trip. The starting time of the trip is randomly selected such that the trip arrives at the school's bus stop 10 to 30 minutes before the school start.

Now our randomly generated county is described by the same kind of data that was provided for the real-world counties. Thus we can proceed along the lines of Section 7.2 to obtain the remaining data describing the instance.

Three random instances rnd_1 to rnd_3 were generated, see Table 7.2. They are much smaller than the real-world instances, having only 25 trips and 10 schools. The instance generator was called three times with exactly the same settings. The number of trips with pupils and the number of transfers are more or less identical. The last two columns show the sizes (i.e., number of variables and constraints) of the integer programming models (4.13) corresponding to the instances. Later we will see that there is a great variety in the time needed to solve these three instance, although they are nearly of the same size.

Chapter 8

Computational Results

We now want to demonstrate how the presented heuristic algorithms for computing schedules and starting times work in practice. The primal and dual algorithms described in the two previous chapters were implemented in C++ (gnu compiler) on a personal computer running Debian Linux as operational system, 2 GByte RAM, and an Intel IV 2.6 GHz CPU. As LP and MIP solver we use CPLEX 9.0 together with ILOG Concert Technology 2.0 (see [42]). We start in Section 8.1 with a very detailed solution of the "tiny" example. Once a feasible schedule of the buses is found, we apply the simple algorithm to compute school and trip starting times. A comparison of the different strategies concerning the greedy and the parametrized greedy (with the three different parameter selection strategies) heuristics is done in Section 8.2 on the set of randomly generated instances. Finally in Section 8.3, we show how the multicriteria model can be used to compute starting times for the real-world instances in a more elaborated way. With this approach, a human planner can interactively take part in the solution finding process by selecting suitable weights for the conflicting goals, until a reasonable solution is found (where "reasonable" depends on the political considerations for the county under consideration).

8.1 Solving the Tiny Instance

In this section we discuss our primal heuristic and some of the dual cutting planes applied to the tiny instance of Section 7.1.

8.1.1 The Primal Greedy Heuristic

In the beginning, the starting time propagation algorithm is called to strengthen the bounds on the time windows.

1. The school time windows do not change, and the trip time windows are narrowed slightly. In particular this means that the instance is feasible.

	s_1	s_2	s_3
$\underline{\tau}_s$	7:30	7:30	7:30
$\overline{\tau}_s$	8:30	8:30	8:30

	t_1	t_2	t_3	t_4	t_5	t_6
$\underline{\alpha}_t$	6:30	6:40	6:35	6:30	6:35	6:30
$\overline{\alpha}_t$	7:55	8:05	8:05	8:10	8:15	8:05

(8.1)

We continue with stage one of the heuristic, the computation of feasible schedules for the buses.

2. We try to eliminate some decision variables. However, no variable can be fixed to its bound. The following tables show the values of the binary variables. Fields with a minus ($-$) entry are non-existing variables (there is no deadhead trip from t_i to t_i). If the field is empty then the corresponding variable is not fixed yet. Otherwise, the entry is either 0 or 1.

	t_1	t_2	t_3	t_4	t_5	t_6
v_t						
w_t						

	t_1	t_2	t_3	t_4	t_5	t_6
t_1	$-$					
t_2		$-$				
t_3			$-$			
t_4				$-$		
t_5					$-$	
t_6						$-$

(8.2)

3. For all deadhead trips $(t_1, t_2) \in \mathcal{A}$, an evaluation of the scoring function $s_{t_1 t_2}$ (with $\lambda_i = 1$ for $i = 1, \dots, 5$) yields the following numbers:

	t_1	t_2	t_3	t_4	t_5	t_6
t_1	$-$	90	75	120	135	120
t_2	90	$-$	90	75	120	105
t_3	75	90	$-$	105	90	105
t_4	140	95	125	$-$	110	110
t_5	135	120	90	90	$-$	90
t_6	180	165	165	150	150	$-$

(8.3)

The minimum score (75) is achieved by deadhead trips $(t_1, t_3), (t_2, t_4)$ and (t_3, t_1), so any of these three is selected. We take the first one to be found, (t_1, t_3). Then we check the consistency of the time windows by calling the starting time propagation algorithm:

	s_1	s_2	s_3
$\underline{\tau}_s$	7:30	7:30	7:30
$\overline{\tau}_s$	8:30	8:30	8:30

	t_1	t_2	t_3	t_4	t_5	t_6
$\underline{\alpha}_t$	6:30	6:40	7:05	6:30	6:45	6:30
$\overline{\alpha}_t$	7:30	7:50	8:05	8:10	8:15	8:05

(8.4)

It is feasible and we can eliminate as many binary variables as possible:

	t_1	t_2	t_3	t_4	t_5	t_6
v_t			0			
w_t	0					

	t_1	t_2	t_3	t_4	t_5	t_6
t_1	–	0	1	0	0	0
t_2		–	0			
t_3	0		–			
t_4			0	–		
t_5				0	–	
t_6	0		0			–

$$(8.5)$$

4. For the remaining deadhead trips, the new scores are computed:

	t_1	t_2	t_3	t_4	t_5	t_6
t_1	–					
t_2	120	–		65	100	100
t_3		130	–	135	110	140
t_4	180	105		–	100	115
t_5	185	140		100	–	105
t_6		170		145	135	–

$$(8.6)$$

Now deadhead trip (t_2, t_4) has the lowest score (65) and therefore will be selected. The resulting time windows are:

	s_1	s_2	s_3
$\underline{\tau}_s$	7:30	7:35	7:30
$\overline{\tau}_s$	8:30	8:30	8:30

	t_1	t_2	t_3	t_4	t_5	t_6
$\underline{\alpha}_t$	6:30	6:40	7:05	7:15	6:35	6:30
$\overline{\alpha}_t$	7:25	7:35	8:05	8:10	8:15	8:05

$$(8.7)$$

More binary variables are fixed to their bounds:

	t_1	t_2	t_3	t_4	t_5	t_6
v_t	1		0	0		
w_t	0	0				

	t_1	t_2	t_3	t_4	t_5	t_6
t_1	–	0	1	0	0	0
t_2	0	–	0	1	0	0
t_3	0	0	–	0		
t_4	0	0	0	–		
t_5	0		0	0	–	
t_6	0	0	0	0		–

$$(8.8)$$

5. The remaining deadhead trips have the following scores:

	t_1	t_2	t_3	t_4	t_5	t_6
t_1	–					
t_2		–				
t_3			–		110	140
t_4				–	145	160
t_5		155			–	105
t_6					135	–

$$(8.9)$$

If we select now deadhead trip (t_5, t_6) with the lowest score (105), then the time windows are no longer feasible. Thus, it is not possible to select (t_5, t_6), and we fix the corresponding variable:

	t_1	t_2	t_3	t_4	t_5	t_6
v_t	1		0	0		
w_t	0	0				

	t_1	t_2	t_3	t_4	t_5	t_6
t_1	−	0	1	0	0	0
t_2	0	−	0	1	0	0
t_3	0	0	−	0		
t_4	0	0	0	−		
t_5	0		0	0	−	0
t_6	0	0	0	0		−

(8.10)

6. Now the scores are:

	t_1	t_2	t_3	t_4	t_5	t_6
t_1	−					
t_2		−				
t_3			−		110	140
t_4				−	145	160
t_5		155			−	
t_6					135	−

(8.11)

The deadhead trip with the lowest score (110) is (t_3, t_5). It is selected, and the resulting time windows are:

	s_1	s_2	s_3
$\underline{\tau}_s$	7:30	7:35	8:00
$\overline{\tau}_s$	8:25	8:30	8:30

	t_1	t_2	t_3	t_4	t_5	t_6
$\underline{\alpha}_t$	6:30	6:40	7:05	7:15	7:45	6:55
$\overline{\alpha}_t$	7:00	7:20	7:35	8:10	8:15	8:05

(8.12)

Then, only a few binary variables remain:

	t_1	t_2	t_3	t_4	t_5	t_6
v_t	1	1	0	0	0	
w_t	0	0	0		1	1

	t_1	t_2	t_3	t_4	t_5	t_6
t_1	−	0	1	0	0	0
t_2	0	−	0	1	0	0
t_3	0	0	−	0	1	0
t_4	0	0	0	−	0	
t_5	0	0	0	0	−	0
t_6	0	0	0	0	0	−

(8.13)

7. Finally, only one deadhead trip remains:

	t_1	t_2	t_3	t_4	t_5	t_6
t_1	−					
t_2		−				
t_3			−			
t_4				−		135
t_5					−	
t_6						−

(8.14)

Thus, deadhead trip (t_4, t_6) is now selected. The new time windows are:

	s_1	s_2	s_3
$\underline{\tau}_s$	7:30	7:30	8:30
$\overline{\tau}_s$	7:45	8:15	8:30

	t_1	t_2	t_3	t_4	t_5	t_6
$\underline{\alpha}_t$	6:30	6:40	7:05	7:15	7:45	8:05
$\overline{\alpha}_t$	6:30	6:40	7:30	7:15	8:15	8:05

(8.15)

Now all decision variables are fixed to their bounds:

	t_1	t_2	t_3	t_4	t_5	t_6
v_t	1	1	0	0	0	0
w_t	0	0	0	0	1	1

	t_1	t_2	t_3	t_4	t_5	t_6
t_1	$-$	0	1	0	0	0
t_2	0	$-$	0	1	0	0
t_3	0	0	$-$	0	1	0
t_4	0	0	0	$-$	0	1
t_5	0	0	0	0	$-$	0
t_6	0	0	0	0	0	$-$

$$(8.16)$$

Thus, the schedules are (t_1, t_3, t_6) and (t_2, t_4, t_5), so the solution only needs two buses instead of six, see Figure 8.1. We apply local search to reduce the length of the deadhead trips, but there is no such possibility.

There is still some flexibility left for the settlement of school and trip starting times. These are computed now in stage two of the heuristic.

8. The starting times of school s_1 can be selected in the interval from 7:30 to 7:45 Since this school currently starts at 8:00, the new starting times is fixed to 7:45. This leads to the following new time windows:

	s_1	s_2	s_3
$\underline{\tau}_s$	7:45	7:35	8:30
$\overline{\tau}_s$	7:45	8:15	8:30

	t_1	t_2	t_3	t_4	t_5	t_6
$\underline{\alpha}_t$	6:30	6:40	7:05	7:15	7:45	8:05
$\overline{\alpha}_t$	6:30	6:40	7:30	7:15	8:15	8:05

$$(8.17)$$

9. The starting time of school s_2 can now be chosen from 7:35 to 8:15. Since this school currently starts at 8:10, the old starting time does not need to be changed. The resulting new time windows are:

	s_1	s_2	s_3
$\underline{\tau}_s$	7:45	8:10	8:30
$\overline{\tau}_s$	7:45	8:10	8:30

	t_1	t_2	t_3	t_4	t_5	t_6
$\underline{\alpha}_t$	6:30	6:40	7:05	7:15	7:45	8:05
$\overline{\alpha}_t$	6:30	6:40	7:30	7:15	8:15	8:05

$$(8.18)$$

10. The starting time of school s_3 is 8:30.

11. The starting time of trip t_1 is 6:30.

12. The starting time of trip t_2 is 6:40.

13. The starting time of trip t_3 is between 7:05 and 7:30. Since this trip currently starts at 7:40, the new starting time is set to 7:30. From this, we derive the following time windows:

	s_1	s_2	s_3
$\underline{\tau}_s$	7:45	8:10	8:30
$\overline{\tau}_s$	7:45	8:10	8:30

	t_1	t_2	t_3	t_4	t_5	t_6
$\underline{\alpha}_t$	6:30	6:40	7:30	7:15	8:10	8:05
$\overline{\alpha}_t$	6:30	6:40	7:30	7:15	8:15	8:05

$$(8.19)$$

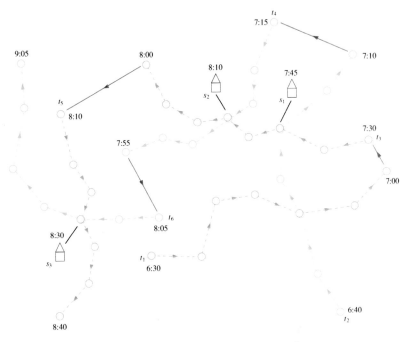

Figure 8.1: The tiny example, optimized

14. The starting time of trip t_4 is 7:15.

15. The starting time of trip t_5 is in the interval from 8:10 to 8:15. Since the current starting time is 7:00, the new starting time will be 8:10. The resulting time windows are:

	s_1	s_2	s_3		t_1	t_2	t_3	t_4	t_5	t_6
$\underline{\tau}_s$	7:45	8:10	8:30	$\underline{\alpha}_t$	6:30	6:40	7:30	7:15	8:10	8:05
$\overline{\tau}_s$	7:45	8:10	8:30	$\overline{\alpha}_t$	6:30	6:40	7:30	7:15	8:10	8:05

(8.20)

16. The starting time of trip t_6 is 8:05.

Summing up, we obtained the following new starting times for the schools: $\tau_{s_1} = 7:45, \tau_{s_2} = 8:10, \tau_{s_3} = 8:30$, i.e., school s_1 starts 15 minutes earlier and school s_3 starts 40 minutes later. The starting time of school s_2 remains unchanged. The starting times of the trips are: $\alpha_{t_1} = 6:30, \alpha_{t_2} = 6:40, \alpha_{t_3} = 7:30, \alpha_{t_4} = 7:15, \alpha_{t_5} = 8:10, \alpha_{t_6} = 8:05$. It is easy to check that this solution satisfies all constraints. Figure 8.1 also shows the new starting times.

8.1.2 Looking at the Dual Side

In this section we prove optimality of the solution found by the heuristic in the previous section. We prove it in three different ways.

Improving The LP Relaxation

Consider the bicriteria model (4.13). We transform it into a single-objective problem by scaling (4.11) by 10000 and (4.12) by 1. Consider then the LP relaxation of model (4.13). This linear program consists of 51 variables and 56 constraints. We solve it to optimality using Dantzig's simplex algorithm. We get 85.1275 as objective function value and the following solution vector x^*, α^*, τ^*. (For simplicity we set $x_{0,t} := v_t$ and $x_{t,0} := w_t$.)

$x^*_{t_i t_j}$	0	t_1	t_2	t_3	t_4	t_5	t_6
0	–	0	0	0	0	0	0
t_1	0	–	0.998	0.002	0	0	0
t_2	0	0	–	0.998	0.002	0	0
t_3	0	0	0.002	–	0.998	0	0
t_4	0	0.9865	0	0	–	0.002	0.0115
t_5	0	0.0115	0	0	0	–	0.9885
t_6	0	0.002	0	0	0	0.998	–

(8.21)

τ^*_s	s_1	s_2	s_3
	7:30	7:55	7:35

α^*_t	t_1	t_2	t_3	t_4	t_5	t_6
	6:30	6:50	7:10	7:35	7:20	6:30

We now tighten the bounds on the starting times and use best possible values for M in inequality (4.8). Solving the LP relaxation gives a significantly improved objective function value 10079.608 (which indicates that we need at least one bus, because this value is greater than 10000) and the following solution:

$x^*_{t_i t_j}$	0	t_1	t_2	t_3	t_4	t_5	t_6
0	–	0	1	0	0	0	0
t_1	0	–	0	0.8339	0	0.1660	0
t_2	0	0	–	0.0335	0.6983	0.2681	0
t_3	0	0	0	–	0.3016	0.2433	0.4549
t_4	0	0.1325	0	0	–	0.3224	0.5450
t_5	0	0.8674	0	0.1325	0	–	0
t_6	1	0	0	0	0	0	–

(8.22)

τ^*_s	s_1	s_2	s_3
	7:50	8:25	7:30

α^*_t	t_1	t_2	t_3	t_4	t_5	t_6
	7:17	7:27	7:33	7:23	6:35	7:05

The boldface numbers (**0** and **1**) in the above table indicate which of the binary variables are already fixed to their lower or upper bounds by (5.19), (6.3), or (6.5). Comparing

(8.21) with (8.22) one can see that improved bounds on the variables and coefficient reduction has a visible influence on the fractional LP optimal solution.

Next on our list is the lifting of coefficients. We include the inequalities (6.21), (6.23) of Theorem 12 and (6.26), (6.29) of Theorem 13 and re-solve the LP relaxation. This yields the objective function value 12819.963 and the following solution:

$x^*_{t_i t_j}$	0	t_1	t_2	t_3	t_4	t_5	t_6
0	—	0.2887	0.6954	0	0.1791	0.1097	0
t_1	0	—	0	0.3834	0	0	0.6166
t_2	0	0	—	0.4410	0.5589	0	0
t_3	0	0.1598	0	—	0.2619	0.1948	0.3834
t_4	0	0	0.3045	0	—	0.6954	0
t_5	0.2730	0.5514	0	0.1755	0	—	0
t_6	1	0	0	0	0	0	—

(8.23)

τ^*_s	s_1	s_2	s_3
	8:15	8:10	8:00

α^*_t	t_1	t_2	t_3	t_4	t_5	t_6
	7:21	7:31	7:19	7:08	7:14	7:33

Comparing (8.22) with (8.23) we see that the starting times of the schools and the trips are now all further away from their lower bounds.

Next we include the valid inequalities of Theorem 14. For the tiny instance, however, they do not lead to any changes in the optimal solution or the objective function value. Thus we turn to the LP-relaxation of the big-M-free formulation (6.58), including the above preprocessing and lifting steps. From the re-computation of the LP optimal solution we obtain an objective function value of 10525.519 and the following solution vector:

$x^*_{t_i t_j}$	0	t_1	t_2	t_3	t_4	t_5	t_6
0	—	0.2893	0.6945	0.0078	0.1797	0.0724	0.0476
t_1	0	—	0	0.3849	0.1846	0	0.4303
t_2	0	0	—	0.4412	0.5587	0	0
t_3	0	0.1676	0	—	0.0768	0.2334	0.5220
t_4	0.0004	0	0.3054	0	—	0.6940	0
t_5	0.2910	0.5430	0	0.1658	0	—	0
t_6	1	0	0	0	0	0	—

(8.24)

τ^*_s	s_1	s_2	s_3
	8:15	8:10	8:00

α^*_t	t_1	t_2	t_3	t_4	t_5	t_6
	7:21	7:31	7:18	7:08	7:14	7:33

For a further improvement of the LP relaxation we now add some cutting planes. We restrict our search for k-path cuts to $k = 2$ and the separation heuristics to subsets with at most 4 elements. That is, we decide by complete enumeration whether there exists a schedule of at most four trips such that 1 vehicle is sufficient to serve them or not. (For this we have to call the starting time propagation at most $4! = 24$ times.) The first separation heuristic cannot identify any cuts. The second heuristic yields the cuts $f(\{2,3,4,5\}) \geq 2, f(\{1,3,5,6\}) \geq 2, f(\{1,3,4,5\}) \geq 2, f(\{1,2,4,5\}) \geq 2$, and the third heuristic returns $f(\{1,2,3,5\}) \geq 2, f(\{1,2,3,4\}) \geq 2$. After including all these inequalities we obtain 16738.030 as objective function value and the corresponding solution

$x^*_{t_i t_j}$	0	t_1	t_2	t_3	t_4	t_5	t_6
0	–	0.6666	0.6666	0.0303	0.3030	0	0
t_1	0	–	0	0.6666	0.3333	0	0
t_2	0	0	–	0.3030	0.3636	0.3333	0
t_3	0	0	0	–	0	0.3636	0.6363
t_4	0	0	0.3333	0	–	0.3030	0.3636
t_5	0.6666	0.3333	0	0	0	–	0
t_6	1	0	0	0	0	0	–

(8.25)

τ^*_s	s_1	s_2	s_3
	7:50	8:05	8:15

α^*_t	t_1	t_2	t_3	t_4	t_5	t_6
	7:05	7:15	7:08	7:03	7:25	7:50

Since we do not find any more k-path cuts we now start with branch-and-bound. We select variable v_{t_1} as branching variable. Setting $v_{t_1} := 1$ gives a feasible integral solution with objective function value 20030. Setting $v_{t_1} := 0$ gives another fractional solution with objective function value 21854.899. Thus the search tree can be pruned, since the dual bound exceeded the best primal bound found before. Thus we have shown optimality of the primal solution of the previous section.

Combinatorial Benders' Cuts

As described in Section 6.1.2 we decompose the model (4.13) for the tiny instance into a master (6.50) and a slave (6.51) problem. The slave is then used to generate combinatorial Benders' cuts, which are iteratively added to the master (6.50).

For the root LP relaxation of the master we get the objective function value 60, which is as expected worse than the objective function value of the LP relaxation of (4.13) (see above). After adding 18 combinatorial Benders' cuts in the root the slightly improved objective function value is 75, and no more cut is found. Then branch-and-cut is activated, where additional cutting planes are generated in every node of the search tree. We generate as many cuts as possible by iteratively searching an IIS (using the findIIS algorithm), adding the corresponding cut, and resolving the current node LP, until no cut is found anymore. After 8 nodes the globally optimal solution with objective function value 20030 is found. Altogether 54 combinatorial Benders' cuts were generated throughout the search. Among them 8 inequalities where of length 1. That is, they are of the form $x_{t_1 t_2} \leq 0$ for some $(t_1, t_2) \in \mathcal{A}$. Hence the corresponding deadhead trip is infeasible, and the variable is fixed to 0. There are 40 inequalities of length 2, i.e., they forbid the simultaneous selection of two deadhead trips. Finally there are 6 inequalities of length 3. (We refrain from listing all 54 inequalities one by one.)

Set Partitioning

Finally we want to demonstrate the set partitioning reformulation of (4.13). For this we have to generate by exhaustive enumeration the column pool $\mathcal{R}$ that consist of all feasible schedules for the problem. This pool has 47 different schedules:

$$
\begin{aligned}
\mathcal{R} := \{ \quad & r_1 := (t_1), & r_2 := (t_2), & \quad r_3 := (t_3), & \quad r_4 := (t_4), \\
& r_5 := (t_5), & r_6 := (t_6), & \quad r_7 := (t_1,t_3,t_4), & \quad r_8 := (t_1,t_3,t_5), \\
& r_9 := (t_1,t_3,t_6), & r_{10} := (t_1,t_3), & \quad r_{11} := (t_1,t_4,t_5), & \quad r_{12} := (t_1,t_4), \\
& r_{13} := (t_1,t_5,t_4), & r_{14} := (t_1,t_5), & \quad r_{15} := (t_1,t_6), & \quad r_{16} := (t_2,t_3,t_4), \\
& r_{17} := (t_2,t_3,t_5), & r_{18} := (t_2,t_3,t_6), & \quad r_{19} := (t_2,t_3), & \quad r_{20} := (t_2,t_4,t_5), \\
& r_{21} := (t_2,t_4,t_6), & r_{22} := (t_2,t_4), & \quad r_{23} := (t_2,t_5,t_4), & \quad r_{24} := (t_2,t_5), \\
& r_{25} := (t_2,t_6), & r_{26} := (t_3,t_4,t_5), & \quad r_{27} := (t_3,t_4), & \quad r_{28} := (t_3,t_5), \\
& r_{29} := (t_3,t_6), & r_{30} := (t_4,t_2,t_5), & \quad r_{31} := (t_4,t_2,t_6), & \quad r_{32} := (t_4,t_2), \\
& r_{33} := (t_4,t_5), & r_{34} := (t_4,t_6), & \quad r_{35} := (t_5,t_1,t_3), & \quad r_{36} := (t_5,t_1), \\
& r_{37} := (t_5,t_2,t_3), & r_{38} := (t_5,t_2,t_4), & \quad r_{39} := (t_5,t_2), & \quad r_{40} := (t_5,t_3,t_4), \\
& r_{41} := (t_5,t_3), & r_{42} := (t_5,t_4,t_2), & \quad r_{43} := (t_5,t_4), & \quad r_{44} := (t_6,t_1), \\
& r_{45} := (t_6,t_2), & r_{46} := (t_6,t_3), & \quad r_{47} := (t_6,t_4) & \quad \}.
\end{aligned}
\tag{8.26}
$$

The objective function values for the respective schedules are

$$
\begin{aligned}
& c_1 := 10000, && c_2 := 10000, && c_3 := 10000, && c_4 := 10000, \\
& c_5 := 10000, && c_6 := 10000, && c_7 := 10020, && c_8 := 10015, \\
& c_9 := 10020, && c_{10} := 10005, && c_{11} := 10030, && c_{12} := 10020, \\
& c_{13} := 10035, && c_{14} := 10025, && c_{15} := 10020, && c_{16} := 10025, \\
& c_{17} := 10020, && c_{18} := 10025, && c_{19} := 10010, && c_{20} := 10015, \\
& c_{21} := 10015, && c_{22} := 10005, && c_{23} := 10030, && c_{24} := 10020, \\
& c_{25} := 10015, && c_{26} := 10025, && c_{27} := 10015, && c_{28} := 10010, \\
& c_{29} := 10015, && c_{30} := 10025, && c_{31} := 10020, && c_{32} := 10005, \\
& c_{33} := 10010, && c_{34} := 10010, && c_{35} := 10030, && c_{36} := 10025, \\
& c_{37} := 10030, && c_{38} := 10025, && c_{39} := 10020, && c_{40} := 10025, \\
& c_{41} := 10010, && c_{42} := 10015, && c_{43} := 10010, && c_{44} := 10020, \\
& c_{45} := 10015, && c_{46} := 10015, && c_{47} := 10010.
\end{aligned}
\tag{8.27}
$$

Initially we let $\mathcal{R}' := \{r_1, \ldots, r_6\}$, that is, we only take those schedules into consideration that consists of one single trip. We solve the LP relaxation (6.61), which yields the following primal and dual solution values:

$$
\begin{aligned}
& y_1 = 1, \quad y_2 = 1, \quad y_3 = 1, \quad y_4 = 1, \quad y_5 = 1, \quad y_6 = 1, \\
& \pi_1 = 1, \quad \pi_2 = 1, \quad \pi_3 = 1, \quad \pi_4 = 1, \quad \pi_5 = 1, \quad \pi_6 = 1,
\end{aligned}
\tag{8.28}
$$

and an objective function value of 60000. The first column (or variable) having negative reduced costs is r_8 (or y_8, respectively). Hence we let $\mathcal{R}' := \mathcal{R}' \cup \{r_8\}$. The new objective function value is 40015 and the corresponding primal and dual solution values are

$$
\begin{aligned}
& y_1 = 0, \quad y_2 = 1, \quad y_3 = 0, \quad y_4 = 1, \quad y_5 = 0, \quad y_6 = 1, \\
& y_8 = 1, \\
& \pi_1 = 1, \quad \pi_2 = 1, \quad \pi_3 = 1, \quad \pi_4 = 1, \quad \pi_5 = 1, \quad \pi_6 = 1,
\end{aligned}
\tag{8.29}
$$

Next column r_{20} enters the pool and we obtain an objective function value of 40015 and

$$
\begin{aligned}
&y_1 = 0, \quad y_2 = 0, \quad y_3 = 0, \quad y_4 = 0, \quad y_5 = 0, \quad y_6 = 1, \\
&y_8 = 1, \quad y_{20} = 1, \\
&\pi_1 = 1, \quad \pi_2 = 1, \quad \pi_3 = 1, \quad \pi_4 = 1, \quad \pi_5 = 0, \quad \pi_6 = 1,
\end{aligned}
\tag{8.30}
$$

Then column r_{21} enters the pool, which gives an objective function value of 30030 and

$$
\begin{aligned}
&y_1 = 0, \quad y_2 = 0, \quad y_3 = 0, \quad y_4 = 0, \quad y_5 = 0, \quad y_6 = 0, \\
&y_8 = 1, \quad y_{20} = 0, \quad y_{21} = 1, \\
&\pi_1 = 1, \quad \pi_2 = 1, \quad \pi_3 = 1, \quad \pi_4 = 0, \quad \pi_5 = 0, \quad \pi_6 = 1,
\end{aligned}
\tag{8.31}
$$

Now column r_9 has negative reduced costs and we obtain 20030 as objective function value (which does not change any more in the sequel) and

$$
\begin{aligned}
&y_1 = 0, \quad\quad y_2 = 0, \quad y_3 = 0, \quad y_4 = 0, \quad\quad y_5 = 0, \quad y_6 = 0, \\
&y_8 = 1, \quad\quad y_{20} = 0, \quad y_{21} = 1, \quad y_9 = 0, \\
&\pi_1 = 0.0015, \quad \pi_2 = 1, \quad \pi_3 = 1, \quad \pi_4 = 0.001, \quad \pi_5 = 0, \quad \pi_6 = 0.0005,
\end{aligned}
\tag{8.32}
$$

as solution vectors. After column r_{19} enters we obtain

$$
\begin{aligned}
&y_1 = 0, \quad y_2 = 0, \quad\quad y_3 = 0, \quad\quad y_4 = 0, \quad\quad y_5 = 0, \quad\quad y_6 = 0, \\
&y_8 = 1, \quad y_{20} = 0, \quad\quad y_{21} = 1, \quad\quad y_9 = 0, \quad\quad y_{19} = 0, \\
&\pi_1 = 1, \quad \pi_2 = 0.9995, \quad \pi_3 = 0.0015, \quad \pi_4 = 0.0015, \quad \pi_5 = 0, \quad \pi_6 = 0.0005.
\end{aligned}
\tag{8.33}
$$

Then column r_7 has negative reduced costs. We get

$$
\begin{aligned}
&y_1 = 0, \quad y_2 = 0, \quad\quad y_3 = 0, \quad\quad y_4 = 0, \quad\quad y_5 = 0, \quad\quad y_6 = 0, \\
&y_8 = 1, \quad y_{20} = 0, \quad\quad y_{21} = 1, \quad\quad y_9 = 0, \quad\quad y_{19} = 0, \quad\quad y_7 = 0, \\
&\pi_1 = 1, \quad \pi_2 = 0.9998, \quad \pi_3 = 0.0012, \quad \pi_4 = 0.0008, \quad \pi_5 = 0.0003, \quad \pi_6 = 0.0008.
\end{aligned}
\tag{8.34}
$$

Finally, column r_{10} enters and we get

$$
\begin{aligned}
&y_1 = 0, \quad y_2 = 0, \quad y_3 = 0, \quad\quad y_4 = 0, \quad\quad y_5 = 0, \quad\quad y_6 = 0, \\
&y_8 = 1, \quad y_{20} = 0, \quad y_{21} = 1, \quad\quad y_9 = 0, \quad\quad y_{19} = 0, \quad\quad y_7 = 0, \quad\quad y_{10} = 0, \\
&\pi_1 = 1, \quad \pi_2 = 1, \quad \pi_3 = 0.0005, \quad \pi_4 = 0.0005, \quad \pi_5 = 0.001, \quad \pi_6 = 0.0015.
\end{aligned}
\tag{8.35}
$$

Now no column has negative reduced costs any more. Since the objective function value of 20030 coincides with the primal solution, we have again shown the optimality of the solution found by the greedy heuristic. Incidentally the optimal solution of the LP relaxation is integral and the schedules corresponding to the selected columns in the final solution (i.e., $y_8 = y_{21} = 1$) are the same, i.e., (t_1, t_3, t_5) and (t_2, t_4, t_6). Note that in general even integral solutions of the set partitioning problems need not correspond to feasible solutions of (4.13), because coupling aspects of the time windows are totally neglected here. Finally only 13 of the originally generated 47 columns were used.

8.2 Solving the Random Instances

In the "tiny" example the number of buses was dramatically reduced from six down to two. As we will see in the sequel for our other random instances the effect is in general not that drastic. In this section we discuss our primal and dual solution approaches for the three randomly generated instances rnd_1 to rnd_3. As underlying model we use the bicriteria formulation (4.13). We turn (4.13) into a single-objective integer program by using weighted sum scalarization. Objective (4.11) is multiplied by 1 and objective (4.12) is multiplied by $\frac{1}{10000}$. This ensures that saving buses is more important than saving deadhead times.

The first thing to do when dealing with a new instance (no matter if random or real-world) is to get an idea how many buses have to be deployed when the trip and school starting times are not changed. We already discussed in Section 3.4 that the remaining problem (i.e., when all starting times are fixed and the optimization solely concentrates on the schedules) is in fact theoretically as well as practically very easy to solve, see Theorem 4. On our computer environment (see above) this takes less than a second. The results for the three instances are given in Table 8.1. For example, in instance rnd_1 there are currently 24 buses used to serve the 25 trips of the instance. Altogether we have 2 minutes for deadhead trips.

Table 8.1: Current number of buses and deadhead trips

instance	objective
rnd_1	24.0002
rnd_2	25.0000
rnd_3	24.0002

For each of the three random instances we specify three different school time windows. In rnd_1a, rnd_2a and rnd_3a, schools may start between 7:45 and 8:15, in rnd_1b, rnd_2b and rnd_3b between 7:30 and 8:30, and in rnd_1c, rnd_2c and rnd_3c between 7:15 and 8:45. Clearly, the wider the school starting time window, the potentially lower the number of deployed vehicles. Altogether we have 9 test instances now.

First we solve them using the primal heuristics. The first heuristic on our list is the classical greedy construction heuristic with scoring function (5.20), once with and once without local search postprocessing. The computational results and times (in seconds) are shown in Table 8.2. The results show that a reduction of the number of deployed buses is achieved by the greedy heuristic. Also the heuristic reacts on the different sizes of the given starting time windows: The larger the windows are, the lower is the number of buses. The length of the deadhead trips was slightly shortened by the local search at the end of the heuristic. The running time for solving each instance is very low, much below one second each.

Next we apply the parametrized greedy heuristic with the scoring function (5.23). In Table 8.3 we compare the final results for pgreedy. Three different strategies were applied for an automated parameter tuning. First we randomly select parameters, then we use hit-and-run, and finally parameters are selected by improving hit-and-run. For all instances we gave an upper limit of 1,000 iterations (i.e., calls of the greedy heuristic with

Table 8.2: Greedy without and with local search

instance	greedy	local search	time
rnd_1a	20.0016	20.0016	< 1
rnd_1b	18.0058	18.0057	< 1
rnd_1c	11.0252	11.0250	< 1
rnd_2a	17.0077	17.0071	< 1
rnd_2b	13.0119	13.0111	< 1
rnd_2c	11.0152	11.0152	< 1
rnd_3a	18.0046	18.0044	< 1
rnd_3b	12.0125	12.0124	< 1
rnd_3c	12.0130	12.0128	< 1

given parameters). After the final iteration local search is applied to the result found by each parameter search, which again slightly lowers the total length of the deadhead trips. The computation times for all three strategies (random, HR, and IHR parameter search) are identical, since the bottleneck is always the call of the greedy heuristic. In all cases, the running time for 1,000 iterations was below 10 seconds. In comparison to the greedy heuristic without parameters (see Table 8.2), pgreedy with any of the three parameter selection strategies is able to detect better solutions (i.e., solutions with a lower objective function value). However, there is no clear winner among the three parameter selection strategies. For each of the three there is an instance where the particular strategy procudes the best result (compared to the two others) after the same number of iterations.

Table 8.3: PGreedy with random, HR, and IHR parameter search

instance	random	HR	IHR	time	iterations
rnd_1a	20.0016	20.0016	20.0016	5	1,000
rnd_1b	13.0304	13.0305	13.0292	7	1,000
rnd_1c	11.0239	11.0239	10.0343	8	1,000
rnd_2a	14.0101	14.0101	14.0101	3	1,000
rnd_2b	11.0144	11.0141	11.0145	3	1,000
rnd_2c	10.0170	10.0168	10.0168	5	1,000
rnd_3a	16.0084	16.0084	16.0083	2	1,000
rnd_3b	11.0186	11.0197	11.0176	2	1,000
rnd_3c	10.0174	10.0174	10.0185	4	1,000

Now we turn to the dual side. We give a time limit of 3,600 seconds (1 hour) for each instance. We tighten the time windows and the values for big-M as described in Section 6.1.1. This is also automatically done by CPLEX when turning on "presolve" (which is already a default setting in CPLEX). First we solve the LP relaxation of model (4.13) for each instance. We remark that the quality of the LP relaxation (in terms of the objective function value) is better the smaller the starting time windows are. The LP relaxation is also the root node of the branch-and-bound tree. Thus in a branch-

and-bound framework there are three benefits from narrow time windows: The number of possible solutions is lower (due to the fact that significantly fewer deadhead trips are available), the dual bound is better, which results in fewer nodes until an optimal solution is found, and finally the LP's are smaller (having less variables and constraints) which makes them faster to solve. The best lower and upper bounds found by the MIP solver is also shown in Table 8.4. If lower and upper bound coincide then a globally optimal solution was found within the given time limit, which happened for 4 of the 9 instances. The integrality gap in these cases is 0%. (We compute this gap as $\frac{u-l}{l}$, where u is the (best) upper bound and l is the (best) lower bound returned by the MIP solver when the time limit is reached.) These results indicate what a modern state-of-the-art commercial solver for mixed-integer programs is able to do for our IOSANA test instances. Compared with the previous table, on these small instances the primal solutions are in nearly all cases better than the solutions found by the pgreedy heuristic. On the other hand, the heuristic was able to find in all but one case solutions with the same number of buses than the MIP solver's primal solutions.

Table 8.4: Dual bounds with standard presolve

instance	var.	cons.	LP relax.	lower bd.	upper bd.	gap	nodes	time
rnd_1a	193	263	10.0006	19.0052	**19.0052**	0%	209	< 1
rnd_1b	475	546	3.3765	13.0292	**13.0292**	0%	51,268	250
rnd_1c	648	719	0.0021	9.4199	10.0343	6.5%	334,020	3,600
rnd_2a	313	373	4.1289	14.0101	**14.0101**	0%	35,191	169
rnd_2b	572	632	0.4047	8.2403	11.0140	33.7%	247,788	3,600
rnd_2c	659	719	0.0016	5.0277	10.0158	99.2%	282,765	3,600
rnd_3a	218	281	7.7555	16.0083	**16.0083**	0%	765	4
rnd_3b	514	604	1.0017	9.5022	11.0178	15.9%	199,157	3,600
rnd_3c	672	735	0.0021	2.6707	10.0161	275.0%	178,594	3,600

There are some binary variables that standard preprocessing is not able to remove from the model, since it would involve preprocessing with more than one inequality at a time (in the sequel called "strong" preprocessing). For general mixed integer preprocessing it would be too time consuming to check every pair or even every triple of inequalities in order to detect some model reductions. Since we have the model formulation at hand we can identify those inequalities more quickly (see Section 6.1.1). The results are given in Table 8.5. As in "normal" preprocessing the benefit within a branch-and-bound framework is threefold. The LPs are smaller and thus faster to solve, the lower bounds are generally better, and the combinatorial complexity is reduced. The previously unsolved instance rnd_1c is now also solved to optimality within the given time limit. For some of the other unsolved instances there are now better lower bounds available, as well in the root node as after the time limit of 3,600 seconds is reached.

Lifting coefficients (see Theorem 11, Theorem 12, and Theorem 13 in Section 6.1.1) has an additional effect on the dual bounds and thus on the branch-and-cut solution process, see Table 8.6. The lower bounds in the root nodes are improved, where the relative

Table 8.5: Dual bounds with strong presolve

instance	var.	cons.	LP relax.	lower bd.	upper bd.	gap	nodes	time
rnd_1a	152	222	10.2810	19.0052	**19.0052**	0%	80	< 1
rnd_1b	413	484	3.3769	13.0292	**13.0292**	0%	5,144	20
rnd_1c	563	634	0.0025	10.0343	**10.0343**	0%	171,100	514
rnd_2a	282	342	4.1294	14.0101	**14.0101**	0%	43,318	151
rnd_2b	520	580	0.6116	9.2908	11.0139	18.5%	654.279	3,600
rnd_2c	590	650	0.0022	5.0031	10.0159	100.2%	325,011	3,600
rnd_3a	189	251	9.0016	16.0083	**16.0083**	0%	537	2
rnd_3b	481	544	1.0021	10.1633	11.0176	8.4%	291,325	3,600
rnd_3c	598	661	0.0026	4.1866	10.0190	139.3%	202,159	3,600

improvement (compared to the corresponding values of Table 8.5) is higher the wider the time windows are. We included in our integer programs the "strong" preprocessing technique, hence the number of variables in the preprocessed integer programs actually is identical. Only the number of inequalities has increased, because previously the trivial bounds (4.1) and (4.2) on the variables α and τ, respectively, were not counted as inequalities.

Table 8.6: Dual bounds with lifting

instance	var.	cons.	LP relax.	lower bd.	upper bd.	gap	nodes	time
rnd_1a	152	310	12.3151	19.0052	**19.0052**	0%	199	< 1
rnd_1b	413	599	5.1029	13.0292	**13.0292**	0%	4,267	40
rnd_1c	563	756	0.9809	10.0343	**10.0343**	0%	224,100	3,511
rnd_2a	282	450	8.3942	14.0101	**14.0101**	0%	7,178	46
rnd_2b	520	704	2.0424	8.2255	11.0140	33.9%	106,681	3,600
rnd_2c	590	774	0.0063	4,6570	10.0163	115.1%	117,500	3,600
rnd_3a	189	333	11.9703	16.0083	**16.0083**	0%	536	3
rnd_3b	481	662	3.7488	9.3286	12.0129	28.8%	155,664	3,600
rnd_3c	598	781	0.0196	4.1569	10.0166	140.9%	79,113	3,600

We now add the CG school cuts of Theorem 14 (see Section 6.1.2) to the problem formulation so far (that is, we keep strong preprocessing and the lifted inequalities). As one can see in Table 8.7 these valid inequalities do not alter the lower bound given by the LP relaxation in the root node. They are however usefull during the branch-and-bound procedure, where for some of the instances (rnd_1a or rnd_2a, for example) a significantly lower number of nodes is needed to find an optimal solution. On the other hand the time for solving the node LPs is higher due to the inclusion of the CG school cuts. Instance rnd_1c is no longer solved to optimality within the given time limit. (A possible way to work around this effect might be the usage of the CG school cut as cutting planes which are only generated on demand to cut off infeasible fractional solutions.)

Table 8.7: Dual bounds with CG school cuts

instance	var.	cons.	LP relax.	lower bd.	upper bd.	gap	nodes	time
rnd_1a	152	461	12.3151	19.0052	**19.0052**	0%	113	< 1
rnd_1b	413	1,253	5.1029	13.0292	**13.0292**	0%	3,440	62
rnd_1c	563	1,704	0.9809	9.5459	10.0343	5.1%	157,175	3,600
rnd_2a	282	813	8.3942	14.0101	**14.0101**	0%	5,151	43
rnd_2b	520	1,546	2.0424	8.6633	11.0145	27.1%	196,100	3,600
rnd_2c	590	1,780	0.0063	5.1045	10.0184	96.2%	125,706	3,600
rnd_3a	189	549	11.9703	16.0083	**16.0083**	0%	141	2
rnd_3b	481	1,399	3.7488	8.9505	12.0135	34.2%	142,529	3,600
rnd_3c	598	1,735	0.0196	3.9423	10.0197	154.1%	75,783	3,600

The next improvement of the model is the use of the big-M-free reformulation of inequalities (4.8) that was presented in Section 6.2.1. In Table 8.8 one can see the additional effect of the big-M-free reformulation on the overall solution process. (Note that the previous changes to the model are kept.) The dual bounds provided by the root LP are once more improved. In particular those instances benefit from the big-M-free reformulation that have the widest time windows. These improvements do not come for free. The price one has to pay for them are linear programs that are larger in the number of variables and constraints. Thus they need more time for solving. After the time limit was reached (for instances that were not solved to optimality before) the number of nodes solved so far is lower than in the other models. On the other hand, the lower bounds from branch-and-bound are better compared to the other models.

Table 8.8: Dual bounds with big-M-free reformulation

instance	var.	cons.	LP relax.	lower bd.	upper bd.	gap	nodes	time
rnd_1a	241	696	12.4504	19.0052	**19.0052**	0%	112	1
rnd_1b	764	2,008	5.5835	13.0292	**13.0292**	0%	3,070	120
rnd_1c	1,066	2,760	2.2844	9.5829	11.0264	15.1%	32,105	3,600
rnd_2a	492	1,295	8.4647	14.0101	**14.0101**	0%	15,737	402
rnd_2b	980	2,517	2.9120	10.0016	11.0138	10.1%	45,600	3,600
rnd_2c	1,119	2,888	1.8844	8.1374	10.0158	23.1%	51,636	3,600
rnd_3a	314	852	11.9809	16.0083	**16.0083**	0%	373	8
rnd_3b	899	2,288	4.0882	10.2826	11.0182	7.1%	46,426	3,600
rnd_3c	1,136	2,861	2.1297	7.6681	10.0162	30.6%	54,800	3,600

The final improvement of the model is the use of k-path cuts. We generate these cuts on demand to strengthen the root relaxation of the model (with all previous changes so far). The further improvements of the dual bound is shown in Table 8.9. We iteratively apply each of the three separation heuristics until all of them fail to produce new inequalities. The objective function value of the model's root LP relaxation was increased by the k-path cuts. The number of inequalities that are found by each of the three heuristics is also given in Table 8.9. Then we use branch-and-bound until either optimality or the time limit is reached. Since the k-path cuts are computationally rather expensive we do not generate them within the search tree. For those instances where optimality was proved within the time limit (such as rnd_2b or rnd3_a) much more nodes are now needed. For other instances, were optimality was not reached within the given time limit, much fewer nodes were solved and the gap between lower and upper bound is larger, compared to the results in Table 8.8. So for our six test instances k-path cuts are only good in improving the root LP, but did not do a good job for branch-and-bound.

Table 8.9: Dual bounds with k-path cuts

instance	cuts (heur. 1,2,3)			LP relax.	lower bd.	upper bd.	gap	nodes	time
rnd_1a	39	13	10	15.9233	19.0052	**19.0052**	0%	61	1
rnd_1b	76	7	5	8.2947	13.0292	**13.0292**	0%	6,590	301
rnd_1c	38	31	53	3.0716	9.2421	10.0343	8.5%	25,832	3,600
rnd_2a	29	4	3	9.7089	14.0101	**14.0101**	0%	11,487	379
rnd_2b	45	10	13	3.8494	9.3260	11.0144	18.1%	29,707	3,600
rnd_2c	42	20	23	2.1046	7.2157	10.0158	38.8%	33,525	3,600
rnd_3a	13	1	2	12.9726	16.0083	**16.0083**	0%	783	11
rnd_3b	10	13	12	4.5758	9.5299	12.0126	26.1%	27,890	3,600
rnd_3c	28	10	9	2.2676	7.0772	10.0199	41.6%	36,350	3,600

The reformulation with combinatorial Benders' cuts was applied to all instances. The results are shown in Table 8.10. The objective function value of the LP relaxation (before adding cutting planes) is here the worst of all (when compared with the previous tables), which is due to the fact that this is only the master problem (6.50) that consits only of inequalities (4.6) and (4.7), and no time window specific constraints. In each node we add as many combinatorial Benders' cuts as possible. The total number of these cuts is given in Table 8.10, whereas in Table 8.11 we give details on the length of the respective cuts.

Table 8.10: Dual bounds with combinatorial Benders' cuts

instance	LP relax.	lower bd.	upper bd.	gap	cuts	nodes	time
rnd_1a	4.0249	19.0052	**19.0052**	0%	485	138	1
rnd_1b	0.0227	12.0318	13.0292	8.3%	7,658	173,700	3,600
rnd_1c	0.0215	3.2294	10.0345	210.7%	13,746	64,200	3,600
rnd_2a	3.0129	14.0101	**14.0101**	0%	2,972	136,900	663
rnd_2b	0.0200	6.0240	11.0149	82.9%	11,594	99,800	3,600
rnd_2c	0.0218	3.0384	10.0212	229.8%	16,831	65,200	3,600
rnd_3a	1.0209	16.0083	**16.0083**	0%	1,203	7,400	21
rnd_3b	0.0215	3.0359	12.0145	295.7%	12,977	72,700	3,600
rnd_3c	0.0213	0.0411	10.0187	22,624.9%	19,224	50,800	3,600

Here the length of a combinatorial Benders' cut is the number of non-zero coefficients. A cut of length zero is in fact a fixation of some binary variable to its lower bound. The majority of all cuts are of length 2 or 3, that is, they are of the form $x_i + x_j \leq 1$ or $x_i + x_j + x_k \leq 2$ for some deadhead trips $i, j, k \in \mathcal{A}$, respectively. Compared to the other approaches above the reformulation based on combinatorial Benders' cut needs much more time to be solved.

Table 8.11: Length of the combinatorial Benders' cuts

instance	1	2	3	4	5	6
rnd_1a	343	141	1	0	0	0
rnd_1b	329	7,098	223	8	0	0
rnd_1c	272	12,760	676	34	4	0
rnd_2a	300	2,594	78	0	0	0
rnd_2b	253	8,405	2,794	140	2	0
rnd_2c	253	7,592	7,339	1,532	114	1
rnd_3a	307	867	29	0	0	0
rnd_3b	256	11,472	1,210	39	0	0
rnd_3c	155	12,792	5,817	438	22	0

We now turn to the set partitioning reformulation. The time for generating the entire pool by enumerating all possible schedules is negligible (less than a second) in all cases. The total number of columns in $\mathcal{R}$ increases when the time windows are wider. The number of columns in the column pool and their length (i.e., number of trips) is shown in Table 8.12.

Table 8.12: The column pools

instance	columns	1	2	3	4	5
rnd_1a	67	25	42	0	0	0
rnd_1b	377	25	227	125	0	0
rnd_1c	863	25	328	485	25	0
rnd_2a	216	25	169	22	0	0
rnd_2b	940	25	319	540	56	0
rnd_2c	3,340	25	347	1,703	1,190	75
rnd_3a	143	25	100	18	0	0
rnd_3b	647	25	338	271	13	0
rnd_3c	2,348	25	445	1,469	406	3

The lower bound from solving the LP relaxation, and the number of columns used in these LPs are shown in Table 8.13. When the time windows are wider also the number of used columns (i.e., the columns which are in $\mathcal{R}'$ when the algorithm terminates) increases, but by far not as fast as the total number of columns. The lower bounds generated by this approach are even better than the previous best lower bounds given by the root LP relaxation presented in Table 8.9. Especially for instances with wider time windows this improvement is quite strong. When the column selection algorithm

Table 8.13: Dual bounds with column generation

instance	col. used	lower bd.	upper bd.	gap	time
rnd_1a	43	17.008	19.0052	11.7%	< 1
rnd_1b	65	10.402	14.0226	34.8%	< 1
rnd_1c	78	10.022	14.0164	39.9%	< 1
rnd_2a	68	11.516	15.0090	30.3%	< 1
rnd_2b	78	9.020	14.0123	55.3%	< 1
rnd_2c	82	9.016	12.0140	33.3%	< 1
rnd_3a	55	13.013	17.0091	30.7%	< 1
rnd_3b	75	8.892	16.0061	80.0%	< 1
rnd_3c	95	7.018	17.0114	142.4%	2

terminates (that is, there are no columns in $\mathcal{R}\backslash\mathcal{R}'$ with negative reduced costs) we compute the best feasible solution for (4.13) that can be built from columns in $\mathcal{R}'$. The results that are given in the column "upper bd." indicate that these solutions are of poor quality, due to the coupled time window aspect that is totally neglected by the column selection process.

Adding combinatorial Benders' cuts, aggregated combinatorial Benders' cuts, set-covering (sc) type cuts, and max clique cuts to the root LP relaxation we can improve both the lower and the upper bounds, see Table 8.14. There are no columns showing the number of combinatorial Benders' cuts and aggregated combinatorial Benders' cuts, because none of those were found. For finding the best possible and feasible combination of schedules from the columns in $\mathcal{R}'$ we make use of the big-M-formulation. Most of the computational time is spent in solving this problem, whereas the time for the column selection is relatively low.

Table 8.14: Dual bounds with column generation and cuts

instance	sc type	max clique	col. used	lower bd.	upper bd.	gap	time
rnd_1a	18	1	48	18.2549	19.0052	4.1%	1
rnd_1b	227	3	144	11.9351	14.0223	17.5%	6
rnd_1c	323	7	235	10.0343	11.0293	9.9%	25
rnd_2a	132	3	114	12.7644	15.0088	17.5%	4
rnd_2b	272	0	195	10.0358	12.0138	19.7%	25
rnd_2c	659	5	511	9.0253	10.0222	11.0%	177
rnd_3a	96	0	79	15.3236	16.0107	4.5%	1
rnd_3b	280	2	181	10.0644	13.0100	29.3%	53
rnd_3c	234	0	215	7.6132	15.0123	97.2%	3,600

Better feasible solutions are found when combining pgreedy with column generation, see Table 8.15. In this hybrid heuristic a limited number of pgreedy runs (100 calls of the greedy heuristic) is used to generate a sufficiently good feasible solution. Then schedules of these solutions are extracted and the corresponding columns with negative reduced

Table 8.15: Primal solutions with pgreedy and column generation

instance	col. used	pgreedy	upper bd.	time
rnd_1a	44	19.0065	19.0052	8
rnd_1b	129	13.0293	13.0292	15
rnd_1c	178	10.0343	10.0343	17
rnd_2a	109	14.0101	14.0101	13
rnd_2b	171	11.0141	11.0138	27
rnd_2c	179	10.0167	10.0159	52
rnd_3a	75	16.0083	16.0083	9
rnd_3b	186	11.0176	11.0176	50
rnd_3c	305	10.0170	10.0163	342

costs are added to $\mathcal{R}'$. After that a column with negative reduced cost is sought. The corresponding deadhead trips of this column are now fixed for the next round of pgreedy iterations. Hence column generation is used as an additional search feature for the pgreedy heuristic. The overall best solution found by this extended pgreedy heuristic is given in column "pgreedy" of Table 8.15. When finally no columns with negative reduced costs can be found in $\mathcal{R}\backslash\mathcal{R}'$ we seek (as in the two tables above) for the best feasible combination of the columns in $\mathcal{R}'$, using the big-M-formulation. Results of this are given in column "upper bd." of Table 8.15.

At the end of this section let us summarize the results from the different tables above. Table 8.16 gives a survey about the increasing lower bounds from the root LP relaxations. The vertical line between the columns "k-path" and "set part." should remind the reader of the change of the underlying model, from constrained network flow to set partitioning.

Table 8.16: The hierarchy of root LP relaxation strengthening

instance	presolve	strong pres.	lifted	big-M-free	k-path	set part.	s.p.&cuts
rnd_1a	10.0006	10.2810	12.3151	12.4504	15.9233	17.008	18.2549
rnd_1b	3.3765	3.3769	5.1029	5.5835	8.2947	10.402	11.9351
rnd_1c	0.0021	0.0025	0.9809	2.2844	3.0716	10.022	10.0343
rnd_2a	4.1289	4.1294	8.3942	8.4647	9.7089	11.516	12.7644
rnd_2b	0.4047	0.6116	2.0424	2.9120	3.8494	9.020	10.0358
rnd_2c	0.0016	0.0022	0.0063	1.8844	2.1046	9.016	9.0253
rnd_3a	7.7555	9.0016	11.9703	11.9809	12.9726	13.013	15.3236
rnd_3b	1.0017	1.0021	3.7488	4.0882	4.5758	8.892	10.0644
rnd_3c	0.0021	0.0026	0.0196	2.1297	2.2676	7.018	7.6132

Summarizing all results from above, we can conclude that the pgreedy heuristic performs reasonable well when compared with the primal solutions found by the MIP solver CPLEX. The quality of dual bounds from the linear programming relaxation highly depends on the size of the time windows: The wider open they are, the more flexibility there is, the worse the bounds. The LP relaxation can be improved by strong pre-processing and by adding problem-specific valid inequalities (cutting planes). However, some of the cuts as well as the big-M-free reformulation have a negative influence on the overall performance of the branch-and-bound process by making the linear programs much more difficult to solve and even increasing the number of subproblems (nodes in the branch-and-bound tree). Thus adding those cuts is not an option to solve larger problem instances. Also combinatorial Benders' cuts did not perform well enough on our random instances. The best lower bounds were in general achieved by solving the LP relaxation of the set partitioning reformulation.

8.3 Solving the Real-World Instances

We now turn from the random to the real-world instances. Most of the calculations can be carried out in the same way as above. Due to the larger size of the instances, however, some of the methods cannot be applied. As a basis we take again the bicriteria model formulation (4.13). As for the random instances, the bicriteria model is turned into a single-criterion problem by scaling objective (4.11) with 1 and objective (4.12) with $\frac{1}{10000}$.

For a comparison with the current number of buses in the respective county we fix the starting time of schools and trips in the model (4.13). Solving the remaining network flow problem yields the results shown in Table 8.17.

Table 8.17: Current number of buses and deadhead trips

instance	objective
Demmin	82.0423
Steinfurt	226.0711
Soest	90.0488
Wernigerode	43.0093
Guetersloh	176.0725

When we started developing a heuristic to solve IOSANA instances we used the nearest neighbor insertion strategy. As one can see from the results shown in Table 8.18 the number of used buses is sometimes even higher than the number of currently used buses. These results stimulated the development of the parametrized greedy heuristic.

Again we test the three different parameter selection strategies (random, hit-and-run, and improving hit-and-run). For each run we give an upper time limit of 3,600 seconds. Hence larger instances will get less iterations than smaller ones. For one data set (Steinfurt) random search found the best solution, for one data set (Guetersloh) hit-and-run was best, and for the remaining three counties improving hit-and-run found the best solution after the same number of iterations.

Table 8.18: Greedy without and with local search

instance	greedy	local search	time
Demmin	81.0362	81.0358	6
Steinfurt	227.0557	227.0556	327
Soest	83.0410	83.0410	11
Wernigerode	49.0138	49.0138	1
Guetersloh	180.0667	180.0663	37

Table 8.19: PGreedy with random, HR, and IHR parameter search

instance	random	HR	IHR	iter.	time
Demmin	65.0953	65.1100	65.0882	2,300	3,600
Steinfurt	173.2122	178.1767	175.1595	120	3,600
Soest	66.1225	66.1377	66.1106	2,500	3,600
Wernigerode	38.0407	38.0417	38.0268	7,500	3,600
Guetersloh	135.1790	133.2179	134.1704	900	3,600

For comparison we now demonstrate how a state-of-the-art MIP solver (CPLEX with default settings) performs within the same time limit. The results are given in Table 8.20.

Table 8.20: Dual bounds, standard presolve

instance	var.	cons.	LP relax.	lower bd.	upper bd.	gap	nodes	time
Demmin	28,566	23,549	15.1263	39.1513	66.0553	68.7%	18,100	3,600
Steinfurt	71,328	65,885	31.6287	88.9567	187.1361	110.4%	240	3,600
Soest	14,620	14,824	6.3873	20.1343	70.0864	248.1%	11,700	3,600
Wernigerode	6,826	4,931	21.3498	37.3984	38.0198	1.6%	174,542	3,600
Guetersloh	57,718	46,406	45.1726	85.6738	140.1084	63.5%	3,451	3,600

Using the "strong" presolve with more than one inequality at a time we can remove some of the binary variables that lead to infeasible solutions when set to their upper bound. The resulting integer programming models are a bit smaller and faster to solve, so that more nodes are computed within a branch-and-bound search. For most instances (except Soest) better lower and upper bounds are found within the time limit. For one real-world instance, Wernigerode, the MIP solver was even able to find a globally optimal solution.

When comparing the best feasible solutions found by the pgreedy heuristic with the best upper bounds from the MIP solver (either using standard or strong preprocessing) it appears that in three of the five cases (Steinfurt, Soest, and Guetersloh) the heuristic (with any of the three parameter selection strategies) returns with better solutions after the same amount of time. For the counties Demmin and Wernigerode the heuristic and

Table 8.21: Dual bounds, strong presolve

instance	var.	cons.	LP relax.	lower bd.	upper bd.	gap	nodes	time
Demmin	28,188	23,170	15.1341	39.1369	65.0585	66.2%	17,211	3,600
Steinfurt	69,992	64,547	31.9827	88.4651	182.1380	105.8%	440	3,600
Soest	14,222	14,426	6.3930	21.5686	73.0879	238.8%	16,700	3,600
Wernigerode	6,768	4,873	23.3905	38.0198	**38.0198**	0%	62,658	1,659
Guetersloh	56,922	45,604	48.7121	86.1151	139.1114	61.5%	3,172	3,600

the MIP solver found feasible solutions with the same number of buses, only the total length of the deadhead trips is shorter for the latter.

We now improve the lower bound using the coefficient lifting techniques, the big-M-reformulation, and the set partitioning root LP relaxation. We do not consider k-path cuts here since solving the root LP relaxation using the big-M-free model already takes up to several hours for the larger instances. As one can see in Table 8.22 the lower bounds are improved, where again the set partitioning reformulation yields the best bounds. The reformulation as a set partitioning problem was only possible for Soest and Wernigerode, because of the large number of possible schedules in the other three cases.

Table 8.22: The hierarchy of root LP relaxation strengthening

instance	presolved	strong pres.	lifted	big-M-free	set part.
Demmin	15.1263	15.1341	22.7273	25.4147	n.a.
Steinfurt	31.6287	31.9827	70.6715	75.1771	n.a.
Soest	6.3873	6.3930	15.0733	18.5776	49.9008
Wernigerode	21.3498	23.3905	28.0122	28.0409	37.3531
Guetersloh	45.1726	48.7121	74.8144	77.1386	n.a.

Combinatorial Benders' cut did not perform very well on our random instances. On the real world instances, the situation is generally the same. Within the given time limit the best upper bound (feasible solution) is bad in general, and the lower bound is also very weak compared to the standard big-M-formulation. Within the given time limit the solver did not finish the root relaxation with combinatorial Benders' cuts for three of the five instances (these three are hence not shown in Table 8.23). However, for the data set Wernigerode these cuts outperform the two other approaches, LP-based branch-and-cut as well as the set partitioning reformulation.

The best lower bounds are found when using the set partitioning reformulation. The values in the "time" column refer to the time that is needed to select columns in $\mathcal{R}'$ from $\mathcal{R}$ so that no column in $\mathcal{R} \backslash \mathcal{R}'$ has negative reduced costs. The time for creating the column pool $\mathcal{R}$ is not taken into cosideration. It took about 48 hours for each of two instances Soest and Wernigerode to enumerate all feasible schedules. For the

Table 8.23: Dual bounds with combinatorial Benders' cuts

instance	LP relax.	lower bd.	upper bd.	gap	cuts	nodes	time
Soest	0.7438	1.9623	165.0157	8,309.3%	11,374	100	3,600
Wernigerode	9.1200	38.0198	**38.0198**	0%	2,558	20,400	401

Table 8.24: Length of the combinatorial Benders' cuts

instance	1	2	3	4	5	6	7	8
Soest	8,496	2,395	286	68	92	30	7	0
Wernigerode	1,780	485	170	84	24	13	1	1

three remaining instances it was not possible to create their respective column pools in reasonable time. Whether it is possible or not depends on the average length of the schedules. Consider for example an instance with 300 trips and an average schedule length (i.e., the number of trips within the schedule) of 3 trips. Then the column pool has about $300^3 = 27$ million entries, which can be done within one week. If the average schedule length would by 4, then we speak of $300^4 = 8,100$ million entries, which would take about 6 years for an exhaustive enumeration. Hence for the instances Demmin, Steinfurt, and Guetersloh the cut pool could not be generated within reasonable time limits.

Table 8.25: The column pools

schedules	Soest	Wernigerode
length 1	191	134
length 2	10,903	5,529
length 3	158,142	94,573
length 4	787,227	697,071
length 5	1,643,006	2,234,949
length 6	1,585,554	3,380,087
length 7	735,043	2,587,369
length 8	156,659	1,015,043
length 9	13,642	195,295
length 10	540	16,950
length 11	0	478
total	5,090,908	10,227,478

Best feasible solutions were again found when combining pgreedy with column selection, see Table 8.27.

Table 8.26: Dual bounds with column generation

instance	col. used	lower bd.	time
Soest	794	49.9008	287
Wernigerode	348	37.3531	277

Table 8.27: Primal solutions with pgreedy and column generation

instance	col. used	pgreedy
Soest	7,479	64.1261
Wernigerode	948	38.0261

Multicriteria Solution Analysis

For the real-world instances the starting time selection problem is a crucial part of the entire planning process. Once improved schedules for the buses are found (i.e., schedules using fewer buses than the current schedules) the starting times can be settled. For this we have two different approaches, both presented in Section 5.2.1. One is the simple first-come-first-serve starting time assignment, the other is the multicriteria assignment using a MIP approach. Once the starting times are selected the objective function values for the remaining goals (i.e., (4.22),..., (4.26)) can be compared with their respective lower and upper bounds. These bounds are computed using a single goal as objective function and neglecting the others. Then a lower bound is obtained by solving a minimization problem, and an upper bound by solving a maximization problem. The results already give interesting insights. The particular upper and lower bounds $\underline{\varepsilon}, \overline{\varepsilon}$ of these objectives are shown in Table 8.28. (Here, $\underline{\varepsilon}_1, \overline{\varepsilon}_1$ are the bounds on goal (4.22), etc.) For example consider the bounds on the waiting time for pupils at their schools $\underline{\varepsilon}_2 \leq \overline{\varepsilon}_2$ in instance Wernigerode. From the lower bound we see that it is not possible to settle the starting time in such way that the pupils have no waiting time at all, even if each trip is served by a new bus! Comparing the upper and the lower bound we see that the waiting time is at most twice the minimum waiting time, but not more. Any feasible solution will be between these two values, and it can be considered as a good solution (for the pupils) if it is closer to the lower than the upper bound.

Table 8.28: Lower and upper bounds on the goals

	$\underline{\varepsilon}_1 \leq \overline{\varepsilon}_1$	$\underline{\varepsilon}_2 \leq \overline{\varepsilon}_2$	$\underline{\varepsilon}_3 \leq \overline{\varepsilon}_3$	$\underline{\varepsilon}_4 \leq \overline{\varepsilon}_4$	$\underline{\varepsilon}_5 \leq \overline{\varepsilon}_5$
Demmin	$0 \leq 13{,}742$	$1{,}845 \leq 5{,}145$	$137 \leq 1{,}244$	$0 \leq 2{,}300$	$0 \leq 8{,}888$
Steinfurt	$0 \leq 15{,}386$	$5{,}281 \leq 10{,}785$	$393 \leq 2{,}321$	$0 \leq 2{,}835$	$0 \leq 10{,}858$
Soest	$0 \leq 6{,}164$	$1{,}994 \leq 5{,}653$	$87 \leq 1{,}439$	$0 \leq 2{,}695$	$0 \leq 6{,}216$
Wernigerode	$0 \leq 6{,}010$	$3{,}110 \leq 6{,}073$	$295 \leq 1{,}123$	$0 \leq 740$	$0 \leq 2{,}680$
Guetersloh	$0 \leq 16{,}741$	$6{,}213 \leq 14{,}500$	$0 \leq 855$	$0 \leq 2{,}190$	$0 \leq 10{,}195$

We also analyze the size of the time windows on the four time variables (starting times and waiting times). For this we carry out the starting time propagation to get tight bounds on all variables. Then we add all upper bounds and subtract all lower bounds (seperately for each of the four variable classes). The results are shown in Table 8.29.

Table 8.29: Size of time windows

	τ	α	ω^{school}	ω^{change}
Demmin	2,890	13,501	4,366	1,650
Steinfurt	4,835	19,057	8,554	4,060
Soest	4,210	9,771	4,468	1,820
Wernigerode	1,310	4,974	7,843	2,040
Guetersloh	3,590	16,735	11,482	1,285

Then we search for a feasible solution using either the MIP solver or the pgreedy heuristic. In the sequel we take the schedules from the pgreedy heuristic with IHR parameter search (see Table 8.19).

The question is how much flexibility remains once the schedules of the vehicles are determined, or vice versa, how much freedom is "lost". To give an answer, Table 8.30 summarizes the size of the time windows after stage one of the greedy heuristic. Comparing the values of Table 8.29 with Table 8.30 one can see that the sizes shrunk by factor 4 to 10.

Table 8.30: Size of time windows for given schedules

	τ	α	ω^{school}	ω^{change}
Demmin	615	3,525	1,902	1,117
Steinfurt	485	2,736	1,884	951
Soest	555	1,076	1,625	606
Wernigerode	645	1,723	2,127	1,069
Guetersloh	625	3,797	2,919	471

We now re-compute the lower and upper bounds on the goals (4.22), ..., (4.26). Consider again the bounds $\underline{\varepsilon}_2 \leq \bar{\varepsilon}_2$ in instance Wernigerode. Due to the fixed schedules in the solution the lower bound has increased from 3,110 to 3,682 and the upper bound has decreased from 6,073 to 5,388.

For the assignment of starting times to the schools and the trips we compare now the two different approaches. First we give in Table 8.32 the results for the simple assignment, where first the starting times are sequentially assigned to the schools and then to the trips, and the starting time propagation is called after each assignment to evaluate the effect on the other starting times. It turns out that the absolute change of the school starting times (4.25) is in all cases at the lower bound. The absolute change of the trip starting times (4.26) is close to the (theoretic) lower bound. The other goals are more or less in the middle between their theoretical lower and upper bounds.

Table 8.31: Lower and upper bounds on the goals for given schedules

	$\varepsilon_1 \le \bar{\varepsilon}_1$	$\varepsilon_2 \le \bar{\varepsilon}_2$	$\varepsilon_3 \le \bar{\varepsilon}_3$	$\varepsilon_4 \le \bar{\varepsilon}_4$	$\varepsilon_5 \le \bar{\varepsilon}_5$
Demmin	$98 \le 1,668$	$3,006 \le 4,607$	$384 \le 1,140$	$275 \le 795$	$1,430 \le 3,682$
Steinfurt	$92 \le 1,299$	$7,251 \le 9,028$	$1,143 \le 1,680$	$2,265 \le 2,665$	$7,839 \le 9,775$
Soest	$115 \le 729$	$3,371 \le 4,829$	$504 \le 962$	$1,110 \le 1,560$	$2,442 \le 3,283$
Wernigerode	$153 \le 1,242$	$3,682 \le 5,388$	$546 \le 986$	$120 \le 545$	$634 \le 1,877$
Guetersloh	$118 \le 2,524$	$9,470 \le 12,232$	$335 \le 692$	$520 \le 985$	$2,580 \le 5,416$

Table 8.32: Results of the simple starting time assignment

	(4.22)	(4.23)	(4.24)	(4.25)	(4.26)
Demmin	536	3,831	730	275	1,531
Steinfurt	269	8,112	1,495	2,265	7,974
Soest	203	4,170	697	1,110	2,468
Wernigerode	490	4,338	704	120	669
Guetersloh	611	11,388	458	520	2,594

In the multicriteria starting time assignment all goals are equally weighted. The objective function values for each goal are listed in Table 8.33. Moreover, comparing these solutions with the bounds given in Table 8.31, one can see that for the majority of the goals they are closer to the lower than to the respective upper bound. Compared to the simple starting time assignment this solution also means greater absolute changes between current and planned starting times of schools and trips. From a practical point of view it is hence questionable whether such solutions can be the basis of negotiations with the school deans.

Table 8.33: Results of the multicriteria starting time assignment

	(4.22)	(4.23)	(4.24)	(4.25)	(4.26)
Demmin	487	3,158	540	330	1,686
Steinfurt	166	7,324	1,229	2,270	8,195
Soest	153	3,540	570	1,200	2,550
Wernigerode	471	3,801	652	280	668
Guetersloh	649	9,659	424	645	2,981

Finally, the objective function values of the above solutions can be compared with the current solution, i.e., the waiting times without optimization (where all schedules and all starting times are as today), see Table 8.34. From this we see that in county Wernigerode the waiting times for the pupils (4.23) in the simple starting time assignment are nearly the same before and after the optimization, and in the multicriteria starting time assignment they are even shorter after the optimization.

All in all, the sum of all waiting times for buses after dead-heading (4.22) is reduced by

Table 8.34: Objective function values for the current solution

	(4.22)	(4.23)	(4.24)	(4.25)	(4.26)
Demmin	2,026	3,550	301	0	0
Steinfurt	1,827	9,550	797	0	0
Soest	592	4,946	404	0	0
Wernigerode	834	4,258	466	0	0
Guetersloh	2,126	11,769	0	0	0

25–90%, and the sum of all waiting times for pupils at their schools (4.23) is reduced by around 10%.

Chapter 9

Concluding Remarks

In this thesis, we started with a real-world planning problem, and translated it into a mathematical language. For the solution of this mathematical problem we proposed mainly two different strategies. On the primal side we proposed a parametrized greedy heuristic. This heuristic is able to find feasible solutions or, as a diagnostic tool, errors in the input data in form of irreducible infeasible subsystems. On the dual side we gave several methods to strengthen the LP relaxation, for example, by preprocessing, cutting planes, or model reformulations.

It turns out that our pgreedy heuristic is able to find good solutions in a very short amount of time. The heuristic can be iteratively used to find school starting times and bus schedules that are also publicly accepted (or at least, not too strongly rejected). The quality of the dual bound of the LP relaxation of the model empirically depends on the size of the time windows. The more narrow the time windows are, the better the bounds and the faster the problem can be solved within a branch-and-cut framework. In the special case of time windows consisting of a single value (i.e., fixed starting times), the problem can even be solved in polynomial time. For problem instances with large time windows the set-partitioning reformulation gives much better dual bounds, if the enumeration of all bus schedules can be carried out in reasonable time.

Besides these solved issues there are several open issues to be addressed in the future. The parametrized greedy heuristic is a new and promising metaheuristic that can be used to solve many other difficult optimization problems. Comparisons with the state-of-the-art heuristics are necessary to estimate its potential. Next, we are still far from routinely solving large problem instances of IOSANA to optimality. For instance, optimal solutions for four of the five real-world instances are still unknown. For this purpose, the development of fast and reliable branch-and-cut algorithms for the solution of set partitioning problems with side constraints is a promising research direction for the future, the cutting planes we presented in this thesis are just the beginning. Another issue is the integration of the (textual) output of the optimization algorithms into a graphical user interface that makes possible real interaction with a human planner.

There is still a lot of interesting and challenging work to do.

What am I waiting for?

Appendix A

IOSANA Models in Zimpl Notation

For rapid mixed-integer programming we use Thorsten Koch's modeling language Zimpl (see [47]).

A.1 The Bicriteria Model

```
# ----------------------------------------------------------------------
# SETS (see Section 4.1)
# ----------------------------------------------------------------------

# bus lines, trips
set V := { read "trips.dat" as "<1n>" };

# deadhead trips
set A := { read "deadheadtrips.dat" as "<1n,2n>" };

# schools
set S := { read "schools.dat" as "<1n>" };

# trips with pupils on board
set P := { read "pupilstrips.dat" as "<1n,2n>" };

# feeder and collector trips
set C := { read "changes.dat" as "<1n,2n>" };


# ----------------------------------------------------------------------
# PARAMETERS (see also Section 4.1)
```

135

```
# --------------------------------------------------------------------

# lower and upper bound on new starting time of school
param tauLower[S] := read "schools.dat" as "<1n> 3n";
param tauUpper[S] := read "schools.dat" as "<1n> 4n";

# trip service duration
param deltaTrip[V] := read "trips.dat" as "<1n> 2n";

# lower and upper bound on new starting time of trip
param alphaLower[V] := read "trips.dat" as "<1n> 3n";
param alphaUpper[V] := read "trips.dat" as "<1n> 4n";

# time for pull-out and pull-in trips
param deltaPullOut[V] := read "trips.dat" as "<1n> 7n";
param deltaPullIn[V]  := read "trips.dat" as "<1n> 8n";

# time from service start for bustrip t to reach school s
param deltaSchool[P] := read "pupilstrips.dat" as "<1n,2n> 3n";

# lower and upper bound on time to walk from bus stop to school
param omegaSchoolLower[P] := read "pupilstrips.dat" as "<1n,2n> 4n";
param omegaSchoolUpper[P] := read "pupilstrips.dat" as "<1n,2n> 5n";

# time from service start for first bustrip ("from") to reach change
busstop
param deltaFeeder[C] := read "changes.dat" as "<1n,2n> 3n";

# time from service start for second bustrip ("to") to reach change
busstop
param deltaCollector[C] := read "changes.dat" as "<1n,2n> 4n";

# time from service start for second bustrip ("to") to reach change
busstop
param omegaChangeLower[C] := read "changes.dat" as "<1n,2n> 5n";
param omegaChangeUpper[C] := read "changes.dat" as "<1n,2n> 6n";

# time for deadhead trip
param deltaShift[A] := read "deadheadtrips.dat" as "<1n,2n> 3n";

# time for waiting after deadhead
param omegaIdleLower[A] := read "deadheadtrips.dat" as "<1n,2n> 4n";
param omegaIdleUpper[A] := read "deadheadtrips.dat" as "<1n,2n> 5n";

# big-M
param M := 10000;
```

```
# -----------------------------------------------------------------------
# VARIABLES AND BOUNDS (see Section 4.2)
# -----------------------------------------------------------------------


# trip t first in block (and new bus required) ?
var v[<t> in V] binary;

# trip t last in block (and send bus back to depot) ?
var w[<t> in V] binary;

# trip t1 connected with t2 ?
var x[<t1,t2> in A] binary;

# (4.1) starting time of bus trip t (and its bounds)
var alpha[<t> in V] integer >=alphaLower[t] <=alphaUpper[t];

# (4.2) starting time of school s (and its bounds)
var tau[<s> in S] integer >=tauLower[s]/5 <=tauUpper[s]/5;


# -----------------------------------------------------------------------
# CONSTRAINTS (see Section 4.3.1)
# -----------------------------------------------------------------------

# (4.6) trip t2 has predecessor or is first in some block
subto predecessor:  forall <t2> in V do
sum <t1,t2> in A do x[t1,t2] + v[t2] == 1;

# (4.7) trip t1 has successor or is last in some block
subto successor:  forall <t1> in V do
sum <t1,t2> in A do x[t1,t2] + w[t1] == 1;

# (4.8) trip starting time bus synchronization for connected trips
subto connectLower:  forall <t1,t2> in A do
alpha[t1] + deltaTrip[t1] + deltaShift[t1,t2] + omegaIdleLower[t1,t2] -
M * (1 - x[t1,t2]) <= alpha[t2];

subto connectUpper:  forall <t1,t2> in A do
alpha[t1] + deltaTrip[t1] + deltaShift[t1,t2] + omegaIdleUpper[t1,t2] +
M * (1 - x[t1,t2]) >= alpha[t2];

# (4.9) bus synchronization for changing pupils
subto changeLower:  forall <t1,t2> in C do
alpha[t2] + deltaCollector[t1,t2] >= alpha[t1] + deltaFeeder[t1,t2] +
```

```
omegaChangeLower[t1,t2];

subto changeUpper:  forall <t1,t2> in C do
alpha[t2] + deltaCollector[t1,t2] <= alpha[t1] + deltaFeeder[t1,t2] +
omegaChangeUpper[t1,t2];

# (4.10) school starts after last trip arrives plus walking time
subto schoolLower:  forall <s,t> in P do
alpha[t] + deltaSchool[s,t] + omegaSchoolLower[s,t] <= 5*tau[s];

subto schoolUpper:  forall <s,t> in P do
alpha[t] + deltaSchool[s,t] + omegaSchoolUpper[s,t] >= 5*tau[s];


# ----------------------------------------------------------------------
# OBJECTIVES (see Section 4.3.2)
# ----------------------------------------------------------------------

minimize objective:

# (4.11) deployed vehicles
sum <t> in V do M * v[t]

# (4.12) pull-out, pull-in and deadhead trips
+ sum <t> in V do deltaPullOut[t] * v[t]
+ sum <t> in V do deltaPullIn[t] * w[t]
+ sum <t1,t2> in A do deltaShift[t1,t2] * x[t1,t2];
```

A.2   The Multicriteria Model

```
# ----------------------------------------------------------------------
# SETS (see Section 4.1)
# ----------------------------------------------------------------------

# bus lines, trips
set V := { read "trips.dat" as "<1n>" };

# deadhead trips
set A := { read "deadheadtrips.dat" as "<1n,2n>" };

# schools
set S := { read "schools.dat" as "<1n>" };
```

```
# trips with pupils on board
set P := { read "pupilstrips.dat" as "<1n,2n>" };

# feeder and collector trips
set C := { read "changes.dat" as "<1n,2n>" };


# ----------------------------------------------------------------------
# PARAMETERS (see also Section 4.1)
# ----------------------------------------------------------------------

# current school starting time
param tauHat[S] := read "schools.dat" as "<1n> 2n";

# lower and upper bound on new starting time of school
param tauLower[S] := read "schools.dat" as "<1n> 3n";
param tauUpper[S] := read "schools.dat" as "<1n> 4n";

# trip service duration
param deltaTrip[V] := read "trips.dat" as "<1n> 2n";

# former starting time of trips
param alphaHat[V] := read "trips.dat" as "<1n> 6n";

# lower and upper bound on new starting time of trip
param alphaLower[V] := read "trips.dat" as "<1n> 3n";
param alphaUpper[V] := read "trips.dat" as "<1n> 4n";

# time for pull-out and pull-in trips
param deltaPullOut[V] := read "trips.dat" as "<1n> 7n";
param deltaPullIn[V] := read "trips.dat" as "<1n> 8n";

# time from service start for bustrip t to reach school s
param deltaSchool[P] := read "pupilstrips.dat" as "<1n,2n> 3n";

# lower and upper bound on time to walk from bus stop to school
param omegaSchoolLower[P] := read "pupilstrips.dat" as "<1n,2n> 4n";
param omegaSchoolUpper[P] := read "pupilstrips.dat" as "<1n,2n> 5n";

# time from service start for first bustrip ("from") to reach change
busstop
param deltaFeeder[C] := read "changes.dat" as "<1n,2n> 3n";

# time from service start for second bustrip ("to") to reach change
busstop
```

```
param deltaCollector[C] := read "changes.dat" as "<1n,2n> 4n";

# time from service start for second bustrip ("to") to reach change
busstop
param omegaChangeLower[C] := read "changes.dat" as "<1n,2n> 5n";
param omegaChangeUpper[C] := read "changes.dat" as "<1n,2n> 6n";

# time for deadhead trip
param deltaShift[A] := read "deadheadtrips.dat" as "<1n,2n> 3n";

# time for waiting after deadhead
param omegaIdleLower[A] := read "deadheadtrips.dat" as "<1n,2n> 4n";
param omegaIdleUpper[A] := read "deadheadtrips.dat" as "<1n,2n> 5n";

# big-M
param M := 10000;


# ------------------------------------------------------------------
# VARIABLES AND BOUNDS (see Section 4.2)
# ------------------------------------------------------------------

# trip t first in block (and new bus required) ?
var v[<t> in V] binary;

# trip t last in block (and send bus back to depot) ?
var w[<t> in V] binary;

# trip t1 connected with t2 ?
var x[<t1,t2> in A] binary;

# (4.1) starting time of bus trip t (and its bounds)
var alpha[<t> in V] integer >=alphaLower[t] <=alphaUpper[t];

# (4.2) starting time of school s (and its bounds)
var tau[<s> in S] integer >=tauLower[s]/5 <=tauUpper[s]/5;

# (4.3) waiting time before school starts (and its bounds)
var omegaSchool[<s,t> in P] real >=omegaSchoolLower[s,t]
<=omegaSchoolUpper[s,t];

# (4.4) waiting time at transfer bus stop (and its bounds)
var omegaChange[<t1,t2> in C] real >=omegaChangeLower[t1,t2]
<=omegaChangeUpper[t1,t2];

# (4.5) idle time after deadhead (and its bounds)
```

```
var omegaIdle[<t1,t2> in A] real >=omegaIdleLower[t1,t2]
<=omegaIdleUpper[t1,t2];

# difference between current and planned school starting time
var DeltaSchool[<s> in S] real >=0;

# difference between current and planned trip starting time
var DeltaTrip[<t> in T] real >=0;

# difference between current and planned pull-out trip t
var DeltaV[<t> in V] binary;

# difference between current and planned pull-in trip t
var DeltaW[<t> in V] binary;

# difference between current and planned deadhead trip from t1 to t2
t
var DeltaX[<t1,t2> in A] binary;


# ----------------------------------------------------------------------
# CONSTRAINTS (see Sections 4.3.1 and 4.4.1)
# ----------------------------------------------------------------------

# (4.6) trip t2 has predecessor or is first in some block
subto predecessor: forall <t2> in V do
sum <t1,t2> in A do x[t1,t2] + v[t2] == 1;

# (4.7) trip t1 has successor or is last in some block
subto successor: forall <t1> in V do
sum <t1,t2> in A do x[t1,t2] + w[t1] == 1;

# (4.14) trip starting time bus synchronization for connected trips
subto connectLower: forall <t1,t2> in A do
alpha[t1] + deltaTrip[t1] + deltaShift[t1,t2] + omegaIdle[t1,t2] - M *
(1 - x[t1,t2]) <= alpha[t2];

subto connectUpper: forall <t1,t2> in A do
alpha[t1] + deltaTrip[t1] + deltaShift[t1,t2] + omegaIdle[t1,t2] + M *
(1 - x[t1,t2]) >= alpha[t2];

# (4.15) bus synchronization for changing pupils
subto changeLower: forall <t1,t2> in C do
alpha[t2] + deltaCollector[t1,t2] == alpha[t1] + deltaFeeder[t1,t2] +
omegaChange[t1,t2];
```

```
# (4.16) school starts after last trip arrives plus walking time
subto schoolLower:  forall <s,t> in P do
alpha[t] + deltaSchool[s,t] + omegaSchool[s,t] == 5*tau[s];


# (4.17) difference between current and planned pull-out trips
subto pullOutDiff1:  forall <t> in V do
vHat[t] - v[t] <= DeltaV[t];


subto pullOutDiff2:  forall <t> in V do
v[t] - vHat[t] <= DeltaV[t];


# (4.18) difference between current and planned pull-in trips
subto pullInDiff1:  forall <t> in V do
wHat[t] - w[t] <= DeltaW[t];


subto pullInDiff2:  forall <t> in V do
w[t] - wHat[t] <= DeltaW[t];


# (4.19) difference between current and planned deadhead trips
subto deadheadDiff1:  forall <t1,t2> in A do
xHat[t1,t2] - x[t1,t2] <= DeltaX[t1,t2];


subto deadheadDiff2:  forall <t1,t2> in A do
x[t1,t2] - xHat[t1,t2] <= DeltaX[t1,t2];


# (4.20) difference between current and planned school starting time
subto schoolDiff1:  forall <s> in S do
tauHat[s] - 5*tau[s] <= DeltaSchool[s];


subto schoolDiff2:  forall <s> in S do
5*tau[s] - tauHat[s] <= DeltaSchool[s];


# (4.21) difference between current and planned trip starting time
subto tripDiff1:  forall <t> in V do
alphaHat[t] - alpha[t] <= DeltaTrip[t];


subto tripDiff2:  forall <t> in V do
alpha[t] - alphaHat[t] <= DeltaTrip[t];



# ------------------------------------------------------------------
# OBJECTIVES (see Sections 4.3.2 and 4.4.2)
# ------------------------------------------------------------------


minimize objective:
```

```
# (4.11) deployed vehicles
sum <t> in V do M * v[t]


# (4.12) pull-out, pull-in and deadhead trips
+ sum <t> in V do deltaPullOut[t] * v[t]
+ sum <t> in V do deltaPullIn[t] * w[t]
+ sum <t1,t2> in A do deltaShift[t1,t2] * x[t1,t2]


# (4.22) idle time after deadhead trips
+ sum <t1,t2> in A do omegaIdle[t1,t2]


# (4.23) waiting time for pupils before school starts
+ sum <s,t> in P do phiSchool[s,t] * omegaSchool[s,t]


# (4.24) waiting time for pupils at transfer bus stops
+ sum <t1,t2> in C do phiChange[t1,t2] * omegaChange[t1,t2]


# (4.25) absolute change of school starting times
+ sum <s> in S do DeltaSchool[s]


# (4.26) absolute change of trip starting times
+ sum <t> in V do DeltaTrip[t]


# (4.27) difference between current and planned schedules
+ sum <t> in V do DeltaV[t]
+ sum <t> in V do DeltaW[t]
+ sum <t1,t2> in A do DeltaX[t1,t2];
```

Bibliography

[1] Ahuja, Magnanti, Orlin (1993), Network Flows: Theory, Algorithms, and Applications. Prentice Hall, New Jersey.

[2] Applegate D., Bixby R., Chvatal V., Cook W. (1995), Finding cuts in the TSP (A preliminary report), DIMACS Technical Report 95-05.

[3] Ascheuer N., Fischetti M., Grötschel M. (2000), A Polyhedral Study of the Asymmetric Traveling Salesman Problem with Time Windows. Networks 36 (2), 69 – 79.

[4] Ascheuer N., Fischetti M., Grötschel M. (2000), Solving the Asymmetric Travelling Salesman Problem with time windows by branch-and-cut. Mathematical Programming 90, 475 – 506.

[5] Aspvell B., Shiloach Y. (1980), A polynomial time algorithm for solving systems of linear inequalities with two variables per inequality. SIAM Journal on Computing 9, 827 – 845.

[6] Balinski M., Quandt R. (1964), On an integer program for a delivery problem. Operations Research 12, 300 – 304.

[7] Bar-Yehuda R., Rawitz D. (2001), Efficient algorithms for integer programs with two variables per constraint. Algorithmica 29 (4), 595 – 609.

[8] Bertsimas D., Tsitsiklis J.N. (1997), Introduction to Linear Optimization. Athena Scientific, Belmont, Massachusetts.

[9] Bierlaire M., Liebling T., Spada M. (2003), Decision-aid Methodology for the School Bus Routing and Scheduling Problem. Conference Paper STRC 2003, 3rd Swiss Transport Research Conference, Ascona.

[10] Bodin L., Berman L. (1979), Routing and Scheduling of School Buses by Computer. Transportation Science 13 (2), 113 – 129.

[11] Bowerman R.L., Hall G.B., Calamai P.H. (1995), A Multi-Objective Optimisation Approach to School Bus Routing Problems. Transportation Research A 28 (5), 107 – 123.

[12] Braca J., Bramel J., Posner B., Simchi-Levi D. (1997), A Computerized Approach to the New York City School Bus Routing Problem. IIE Transactions 29, 693 – 702.

[13] Bramel J., Simchi-Levi D. (2002), Set-Covering-Based Algorithms for the Capacitated VRP. In: Toth P., Vigo D. (eds.), The vehicle routing problem. SIAM Monographs on Discrete Mathematics and Applications. SIAM, Philadelphia, 85 – 108.

[14] Bruni R. (2003), On Exact Selection of Minimally Unsatisfiable Subformulae. Annals of Mathematics and Artificial Intelligence 43, 35-50.

[15] Chinneck J.W., Dravnieks E.W. (1991), Locating minimal infeasible constraint sets in linear programming. ORSA Journal on Computing 3, 157 – 168.

[16] Chvatal V. (1973), Edmonds Polytopes and a Hierarchy of Combinatorial Problem. Discrete Mathematics 4, 305 – 337.

[17] Clark J., Holton D. (1994), Graphentheorie. Spektrum Akademischer Verlag, Heidelberg. (In German).

[18] Clark G., Wright J.V. (1964), Scheduling of vehicles from a central depot to a number of delivery points. Operations Research 12, 568 – 581.

[19] Cohen E., Megiddo N. (1994), Improved algorithms for linear inequalities with two variables per inequality. SIAM Journal on Computing 23, 1313 – 1347.

[20] Codato G., Fischetti M. (2004), Combinatorial Benders' Cuts. In: Bienstock D., Nemhauser, G. (eds.), IPCO 2004, LNCS 3064, Springer Verlag Berlin Heidelberg, 178 – 195.

[21] Corberan A., Fernandez E., Laguna M., Marti R. (2000), Heuristic Solutions to the Problem of Routing School Buses with Multiple Objectives. Technical Report TR08-2000, Dep. of Statistics and OR, University of Valencia, Spain.

[22] Cordeau J.-F., Desaulniers G., Desrosiers J., Solomon M.M., Soumis F. (2002), VRP with time windows. In: Toth P., Vigo D. (eds.), The vehicle routing problem. SIAM Monographs on Discrete Mathematics and Applications. SIAM, Philadelphia, 157 – 193.

[23] Dantzig G.B., Fulkerson D.R., Johnson S. (1954), Solution of a Large Scale Traveling Salesman Problem. Journal of the Operations Research Society of America 2, 393 – 410.

[24] Dantzig G.B., Ramser J.H. (1959), The truck dispatching problem. Management Science 6, 80.

[25] Desaulniers G., Desrosiers J., Erdmann A., Solomon M.M., Soumis F. (2002), VRP with pickup and delivery. In: Toth P., Vigo D. (eds.), The vehicle routing problem. SIAM Monographs on Discrete Mathematics and Applications. SIAM, Philadelphia, 225 – 242.

[26] Desaulniers G., Lavigne J., Soumis F. (1998), Multi-Depot Vehicle Scheduling with Time Windows and Waiting Costs. European Journal of Operational Research 111, 479 – 494.

[27] Desrochers M., Laporte G. (1991), Improvements and extensions to the Miller-Tucker-Zemlin subtour elimination constraints. Operations Research Letters 10, 27 – 36.

[28] Dijksta E.W. (1959), A Note on Two Problems in Connexion with Graphs. Numer. Math. 1, 269 – 271.

[29] Ehrgott M. (2000), Multicriteria Optimization. Springer Verlag, Heidelberg.

[30] Fourer R. (2003), Linear Programming Software Survey. OR/MS Today, Lionheart Publishing, Georgia.

[31] Fügenschuh A., Martin A. (2004), Verfahren und Vorrichtung zur automatischen Optimierung von Schulanfangszeiten und des öffentlichen Personenverkehrs und entsprechendes Computerprogramm. Deutsche Patentanmeldung 10 2004 020 786.0. (In German.)

[32] Fügenschuh A., Martin A. (2005), Computational Integer Programming and Cutting Planes, K. Aardal, G. Nemhauser, R. Weissmantel (Hrsg.), "Handbook on Discrete Optimization", Series "Handbooks in Operations Research and Management Science".

[33] Fulkerson D.R. (1971), Blocking and Anti-Blocking Pairs of Polyhedra. Mathematical Programming 1, 168 – 194.

[34] Garey M.R., Johnson D.S. (1979), Computers and Intractability: A Guide to the Theory of NP-Completeness. W.H. Freeman and Company, New York.

[35] Ginter V., Kliewer N., Suhl L. (2005), Solving large multi-depot multi-vehicle-type bus scheduling problems in practice. OR Spectrum 27, 507 – 523.

[36] Gomory R.E. (1958), Outline of an Algorithm for Integer Solutions to Linear Programs. Bulletin of The American Mathematical Society 64, 275 – 278.

[37] Grötschel M., Lovász L., Schrijver A. (1988), Geometric Algorithms and Combinatorial Optimization. Springer Verlag, New York.

[38] Guieu O., Chinneck J.W. (1999), Analyzing Infeasible Mixed-Integer and Integer Linear Programs. INFORMS Journal on Computing 11(1), 63 – 77.

[39] Hinweise zur Schulorganisation für allgemein bildende Schulen. Verwaltungsvorschrift des Ministeriums für Bildung, Wissenschaft und Kultur vom 21. Juli 2000. Aus: Mitteilungsblatt des Ministeriums für Bildung, Wissenschaft und Kultur Mecklenburg-Vorpommern, Nr. 8/2000, 362 – 365. (In German).

[40] Hochbaum D.S., Megiddo N., Naor J., Tamir A. (1993), Tight bounds and 2-approximation algorithms for integer programs with two variables per inequality. Mathematical Programming 62, 69 – 83.

[41] Hochbaum D.S., Naor J. (1994), Simple and fast algorithms for linear and integer programs with two variables per inequality. SIAM Journal on Computing 23 (6), 1179 – 1192.

[42] ILOG CPLEX Division, 889 Alder Avenue, Suite 200, Incline Village, NV 89451, USA. Information available at URL http://www.cplex.com.

[43] Kaballo W. (1996), Einführung in die Analysis I. Spektrum Akademischer Verlag, Heidelberg. (In German).

[44] Kaballo W. (1998), Einführung in die Analysis III. Spektrum Akademischer Verlag, Heidelberg. (In German).

[45] Kara I., Laporte G., Bektas T. (2003), A Note on the Lifted Miller-Tucker-Zemlin Subtour Elimination Constraints for the Capacitated Vehicle Routing Problem. Technical Report, Les Cahiers du GERAD, G-2003-12.

[46] Keller H., Müller W. (1979), Optimierung des Schülerverkehrs durch gemischt ganzzahlige Programmierung. Zeitschrift für Operations Research B 23, 105 – 122. (In German).

[47] Koch T. (2004), Rapid Mathematical Programming. PhD Thesis, Berlin.

[48] Kohl N., Desrosiers J., Madsen O.B.G., Solomon M.M., Soumis F. (1999), 2-Path cuts for the vehicle routing problem with time windows. Transportation Science 33, 101 – 116.

[49] Lagarias J.C (1985), The computational complexity of simultaneous diophantine approximation problems. SIAM Journal on Computing 14, 196 – 209.

[50] Land A.H., Doig A.G. (1960), An Automatic Method for Solving Discrete Programming Problems. Econometrica 28, 497 – 520.

[51] Laporte G., Semet F. (2002), Classical heuristics for the capacitated VRP. In: Toth P., Vigo D. (eds.), The vehicle routing problem. SIAM Monographs on Discrete Mathematics and Applications. SIAM, Philadelphia, 109 – 128.

[52] Löbel A. (1997), Optimal Vehicle Scheduling in Public Transit. Shaker Verlag, Aachen.

[53] Maffioli F., Sciomachen A. (1997), A mixed-integer model for solving ordering problems with side constraints. Annals of Operations Research 69, 277 – 297.

[54] Martin A. (1999), Integer Programs with Block Structure. Habilitationsschrift, Technische Universität Berlin.

[55] Megiddo N. (1983), Towards a genuinely polynomial algorithm for linear programming. SIAM J. Comput. 12, 347 – 353.

[56] Miller C.E., Tucker A.W., Zemlin R.A. (1960), Integer Programming Formulation of Traveling Salesman Problems. Journal of the ACM 7, 326 – 329.

[57] Müller-Merbach H. (1983), Zweimal Travelling Salesman, DGOR-Bulletin 25, 12 – 13. (In German).

[58] Nelson C.G. (1978), An $n^{\log n}$ algorithm for the two-variable-per-constraint linear programming satisfiability problem. Technical Report AIM-319, Stanfort University.

[59] Nemhauser G., Wolsey L. (1988), Integer and Combinatorial Optimization. Wiley-Interscience. John Wiley & Sons, New York.

[60] Padberg M.W. (1973), On the Facial Structure of Set Packing Polyhedra. Mathematical Programming 5, 199 – 215.

[61] Padberg M.W. (1975), A Note on Zero-One Programming. Operations Research 23 (4), 833 – 837.

[62] Pratt V.R. (1977), Two easy theories whose combination is hard. Technical Report, Massachusetts Institute of Technology. Cambridge, Massachusetts.

[63] Savelsbergh M. (1986), Local Search for Routing Problems with Time Windows. Annals of Operations Research 4, 285 – 305.

[64] Schrijver A. (1980), On cutting planes. Annals of Discrete Mathematics 9, 291 – 296.

[65] Schrijver A. (2005), On the history of combinatorial optimization (till 1960). Technical Report, Center for Mathematics and Computer Science, Amsterdam. Online available at URL http://homepages.cwi.nl/~lex

[66] Satzung über die Anerkennung der notwendigen Kosten für die Schüler-beförderung des Landkreises Demmin. Aus: Kreisanzeiger des Landkreises Demmin, Nr. 6/97. Amtliche Mitteilungen und Informationen, 7 – 8. (In German).

[67] Schulgesetz Mecklenburg-Vorpommern vom 15.05.1996. Stand Feb. 2000, 68. (In German).

[68] Shostak R. (1981), Deciding linear inequalities by computing loop residues. Journal of the ACM 28, 769 – 779.

[69] Stöveken P. (2000), Wirtschaftlicherer Schulverkehr: ÖPNV-Optimierung mit erfolgsabhängiger Honorierung. Der Nahverkehr 3, 65 – 68. (In German).

[70] Tamiz M., Mardle S.J., Jones D.F. (1996), Detecting IIS in Infeasible Linear Programmes Using Techniques from Goal Programming. Computers in Operations Research 23(2), 113 – 119.

[71] Toth P., Vigo D. (2002), The Vehicle Routing Problem. SIAM Monographs on Discrete Mathematics and Applications. SIAM, Philadelphia.

[72] Toth P., Vigo D. (2002), VRP with backhauls. In: Toth P., Vigo D. (eds.), The vehicle routing problem. SIAM Monographs on Discrete Mathematics and Applications. SIAM, Philadelphia, 195 – 224.

[73] van Eijl C.A. (1995), A polyhedral approach to the delivery man problem. Technical Report 95-19. Department of Mathematics and Computer Science, Eindhoven University of Technology.

[74] van Loon J.N.M. (1981), Irreducible inconsistent systems of linear inequalities. European Journal of Operations Research 8, 283 – 288.

[75] Wikipedia. Online available at URL http://www.wikipedia.org

[76] Wolsey L.A. (1975), Faces of Linear Inequalities in 0-1 Variables. Mathematical Programming 8, 165 – 178.

[77] Wolsey L.A. (1998), Integer Programming. John Wiley & Sons, New York.

[78] Zabinsky Z.B. (2003), Stochastic Adaptive Search for Global Optimization. Nonconvex Optimization and its Applications. Kluwer Academic Publishers, Boston.

[79] Zabinsky Z.B., Smith R.L., McDonald J.F., Romeijn H.E., Kaufman D.E. (1993), Improving Hit-and-Run for Global Optimization. Journal of Global Optimization 3, 171 – 192.

Curriculum Vitae

Armin Fügenschuh

geboren am 28. September 1974 in Cuxhaven.

1980 - 1984:	Grundschule in Böblingen
1984:	Grundschule in Aurich
1984 - 1986:	Orientierungsstufe in Aurich
1986 - 1994:	Gymnasium in Aurich
8. Juni 1994:	Allgemeine Hochschulreife
1994 - 1995:	Grundwehrdienst
1995 - 2000:	Studium der Mathematik und Informatik an der Carl von Ossietzky Universität in Oldenburg
30. Sep. 2000:	Diplom in Mathematik
seit 2000:	Wissenschaftlicher Mitarbeiter an der Technischen Universität in Darmstadt, Fachbereich Mathematik, Arbeitgruppe Diskrete Optimierung